INTRODUCTION TO ERGODIC THEORY

YA. G. SINAI

TRANSLATED BY V. SCHEFFER

PRINCETON UNIVERSITY PRESS

1976

Published by Princeton University Press, Princeton, New Jersey

In the United Kingdom: Princeton University Press, Chicester, West Sussex

Library of Congress Cataloging-in-Publication Data

Sinaĭ, ÎAkov Grigor'evich, 1935-
Introduction to ergodic theory.

(Mathematical notes ; 18)
Based on a series of lectures given at the Moscow and Erevan State Universities.
1. Ergodic theory. I. Title. II. Series: Mathematical notes (Princeton, N. J.) ; 18.
QA313.S56 515'.42 76-3030
ISBN 0-691-08182-4

This book was originally published in Russian by the Erevan State University, U.S.S.R., 1973

Princeton University Press books are printed on acid-free paper and meet the guidelines for permanence and durability of the Committee on Production Guidelines for Book Longevity of the Council on Library Resources.

http://pup.princeton.edu

Printed in the United States of America

Table of Contents

Introduction to Ergodic Theory

Ya. G. Sinai

Introduction.

The expression "exposition for pedestrians" is often encountered among physicists. This is usually understood to mean the exposition of some theory that is destined for a wide circle of readers in which the fundamental ideas are moved to the foreground and the technical results and more refined questions are ignored. In this sense these lectures may be called an exposition of ergodic theory "for pedestrians". They originated from a series of lectures which I delivered to students of the third year, and higher, in the mechanical-mathematical faculty of the Moscow and Erevan State Universities. In this work the reader will seldom find proofs of general theorems based on a serious application of measure theory. On the other hand, a large number of examples that are popular in ergodic theory is examined. The knowledge of the most general facts of measure theory of the theory of smooth manifolds, and of probability theory is sufficient to understand them. It must be noted, by the way, that in a number of respects the development of the theory has not gone much further than the investigation of these examples.

The proofs in this course are not always carried out to the end, for an accurate construction of them is not possible on the level (adopted) here.

The references to the literature which are found in the text should help the reader familiarize himself with the deeper literature on the questions discussed. We emphasize that these references are not complete.

The last two lectures are dedicated to entropy. For their understanding it is helpful to know the theory of measurable decompositions and, in particular, operations on decompositions.

The students A. Brudno and S. Zenovich of the mechanical-mathematical faculty of M.S.U. rendered great help during the preparation of these lectures for publication, and I present my sincere thanks to them.

Lecture 1.

THE FUNDAMENTAL PROBLEMS IN ERGODIC THEORY

What are the basic problems in ergodic theory? From my point of view, the basic problems in ergodic theory consist of the study of the statistical properties of the groups of motions of non-random objects. It must be emphasized that this is my point of view, and that other points of view are completely valid. For example, it may be considered that ergodic theory studies the category of measure spaces in which the morphisms are the measure preserving transformations.

Now we will briefly define what we will understand by statistical properties of groups of motions. The basic space will always be denoted $\mathfrak{M}$. It is clear that $\mathfrak{M}$ must be a measurable space; i.e. a certain σ-algebra γ of subsets must be defined on it. In all concrete cases the determination of this σ-algebra presents no difficulties. We suppose that a certain group or semigroup of transformations G acts on $\mathfrak{M}$. For the time being we will examine the case where G is countable. Then for any element $g \in G$ the transformation $T_g : \mathfrak{M} \to \mathfrak{M}$ is defined so that

1) T_g is a measurable transformation; i.e. if $A \in \gamma$ then $T_g(A)$, $T_g^{-1}(A) \in \gamma$;

2) $T_{g_1} \circ T_{g_2} = T_{g_1 g_2}$.

It follows from 2) that, in the case where G is a group, every T_g is invertible, for

$$(T_g)^{-1} = T_{g^{-1}} .$$

The following examples will be useful in the future.

Example 1. $G = \mathbb{Z}^+$, $T = T_1$ the translation by one unit, is an endomorphism of the space $\mathfrak{M}$; i.e. a single valued but not one-to-one transformation on $\mathfrak{M}$.

Example 2. $G = \mathbb{Z}$, $T = T_1$ is a bijection on the space $\mathfrak{M}$, called an automorphism, and $T_n = (T_1)^n = T^n$.

It is possible to examine a more general situation, in which G is an arbitrary countable commutative group. But for the time being we will restrict ourselves to the two examples. Now we will touch on an extremely important and fairly general problem:

What does it mean to say that the action of the group G has random (or statistical) properties?

The scheme that is proposed below is not the most general one. For instance, it does not include the application of ergodic theory to the theory of algebraic fields which is dealt with in the book "Ergodic Theory of Algebraic Fields" by U. V. Linnik.

A certain accumulation of experience gives basis to the introduction of the following five related properties which to a certain extent reinforce each other and which may naturally be called statistical.

Property I. The existence of a finite measure μ that is invariant under G. (It is always possible to obtain $\mu(\mathfrak{M}) = 1$ by normalization).

The invariance of the measure with respect to the action of the group G means that for any set $A \in \gamma$ and for any element $g \in G$

$$\mu(A) = \mu(T_g^{-1}A) .$$

Since $\mathcal{M}$ is a measurable space, it is possible to examine the measurable functions (random variables) f(x) and the corresponding adjoint (i.e. acting on functions) semigroup or group of transformations $\{\mathcal{U}_g\}$

$$\mathcal{U}_g : f(x) \to f(T_g x) \; ; \; G \to \{\mathcal{U}_g\} .$$

In that case the invariance of the measure μ is equivalent to the following relationship:

$$\int f(x) d\mu(x) = \int (\mathcal{U}_g f)(x) d\mu(x) \qquad (1)$$

that must be satisfied for every $g \in G$.

We will prove this assertion. By virtue of the linearity of mathematical expectation it suffices to verify (1) for characteristic functions of measurable sets.

The function

$$\chi_A(x) = \begin{cases} 1, & \text{if } x \in A \\ 0, & \text{if } x \notin A \end{cases}$$

is called the characteristic function of the set A. We set $f(x) = \chi_A(x)$. Substituting into (1), we obtain

$$\mu(A) = \int_{\mathcal{M}} \chi_A(x)d\mu(x) \; ; \int_{\mathcal{M}} (\mathcal{U}_g f)(x)d\mu(x)$$

$$= \int_{\mathcal{M}} \chi_{T_g^{-1}A}(x)d\mu(x) = \mu(T_g^{-1}A) \, ,$$

and we reach the original definition of invariance.

If G is a group then $T_g^{-1} = T_{g^{-1}}$ and hence $\mu(A) = \mu(T_g^{-1}A) = \mu(T_g A)$; i.e. the measure of any measurable set equals the measure of the image and the inverse image of that set.

The next theorem shows why the presence of an invariant measure can be related to the statistical properties of the action of the group G. In its formulation G will be either $\mathbb{Z}^+$ or $\mathbb{Z}$, even though a more general situation may be examined.

Ergodic Theorem of Birkhoff-Khinchin.

Let $f(x) \in \mathcal{L}_\mu^1(\mathcal{M})$ and let μ be invariant with respect to G. Then the following limit as $n \to \infty$ exists with probability 1:

$$\lim_{n\to\infty} \frac{1}{n} \sum_{k=0}^{n-1} f(T^k x) = \lim_{n\to a} \frac{1}{n} \sum_{k=0}^{n-1} \mathcal{U}^a f = \hat{f}(x) \, .$$

It also follows that $\hat{f}(Tx) = \hat{f}(x)$ with probability 1 and

$$\int_{\mathcal{M}} f(x)d\mu(x) = \int_{\mathcal{M}} \hat{f}(x)d\mu(x).$$

In case G is a group we have

$$\lim_{n\to\infty} \frac{1}{n}\sum_{k=0}^{n-1} f(T^k x) = \lim \frac{1}{n}\sum_{k=0}^{n-1} f(T^{-k}x)$$

with probability 1.

In the theory of probability similar assertions are called laws of large numbers; and since the convergence takes place almost everywhere the ergodic theorem of Birkhoff and Khinchin is a theorem of the type of a strangthened law of large numbers. The proof of this theorem can be found in the book "Ergodic Theory and Information" by P. Billingsley.

Very often, in particular in the physical literature, it is not sufficiently emphasized that the ergodic theorem of Birkhoff-Khinchin holds almost everywhere. Those points x for which it holds in the case of good functions can be naturally called typical. Nontypical points may be encountered in real situations, and that complicates research considerably.

The fact that in the presence of an invariant measure it is possible to take averages over time, implies that the system is in a stationary condition, unchanging over time. It often occurs that a system approaches some stationary regime with the elapse of time. Then it is natural to study that regime with the aid of the corresponding invariant measure.

Now, we will obtain the first and the simplest consequence of the existence of an invariant measure.

Poincaré's Recurrence Theorem.

Let the group $\{T^n\}$ of powers of an automorphism T act on the space $\mathfrak{M}$. For every set A such that $\mu(A) > 0$ and for almost every point $x \in A$ there exists an infinite increasing sequence of numbers n_i for which $T^{n_i}x \in A$.

Before proving this theorem I will describe the Zermelo paradox, which is related to the Poincaré recurrence theorem.

We will study a gas enclosed in a finite volume. (Jumping ahead, we remark that the Poincaré theorem is valid also for continuous time $G = R^1$. The proof extends with little change).

In classical statistical mechanics a gas is regarded as composed of a large number of molecules, interacting with each other and moving according to the laws of classical mechanics. In other words, such a gas is a hamiltonian system, but a system with a very large number of degrees of freedom (under normal conditions, a cubic centimeter contains some 10^{20} molecules). As we will see later, every closed hamiltonian system in a finite volume has a finite invariant measure and the Poincaré theorem on recurrence applies. We will take for the set A the collection of those initial data for which all molecules are found in the left half of the container. It is easy to see from the form of the measure that $\mu(A) > 0$. Then it follows from the Poincaré recurrence theorem that for almost every point $x \in A$ there must exist an arbitrarily large number of moments in time when the trajectory of the point X is in the set A. However, not a single incident has been recorded in all of human history when all the molecules of a gas returned to occupy half of their container. This paradox is called Zermelo's paradox and is connected with the foundations of statistical mechanics. For its resolution it is commonly said that the cycles of Poincaré are so long that they exceed the lifespan of the galaxy and, in particular, the lifespan of the gas container.

We proceed to the proof of Poincaré's Recurrence theorem.

We denote by A_1 the set of points in $\mathcal{A}$ which return to $\mathcal{A}$ at least once: $A_1 = \{x : x \in A, \exists k > 0, T^k x \in A\}$.

If we prove that $\mu(A) = \mu(A_1)$ then the assertion of the theorem easily follows. Indeed, let $A_1^{(k)} = \{x : x \in A_1$ and $k > 0$ is the smallest number for which $T^k x \in \mathcal{A}\}$, $\mathcal{B}_1^{(k)} = T^k A_1^{(k)}$. Then $\mathcal{B}_1^{(k_1)} \cap \mathcal{B}_1^{(k_2)} = \phi$, $\mathcal{B}_1^{(k)} \subset A$ and $\sum_k \mu(\mathcal{B}_1^{(k)}) = \sum_k \mu(A_1^{(k)}) = \mu(A_1) = \mu(A)$. Let us denote $A_2 = \{x : x \in A_1^{(k)}$ and $T^k x \in A_1\} = \mathcal{A}_1 \cap (\bigcup_{k>0} \mathcal{B}_1^{(k)})$. The set A_2 consists of $x \in A$ which return in A at least twice. We have $\mu(A_2) = \mu(\mathcal{A})$ because $\mu(\bigcup_k \mathcal{B}_1^{(k)}) = \sum_n \mu(\mathcal{B}_1^{(k)}) = \mu(A)$ and so forth. If we denote by A_i the subset of A consisting of $x \in A$, which return to A at least i times then we shall have in analogues way $\mu(A_i) = \mu(A)$ and therefore $\mu(\bigcap_i \mathcal{A}_i) = \lim_{i\to\infty} \mu(A_i) = \mu(A)$.

We set $C = A \setminus A_1 = \{x : x \in A, T^k x \notin A$ for all $k > 0\}$ Assume that $\mu(C) > 0$. We have $T^{-k} C \cap C = \phi$ for all k. That implies that for all $k > 0$ the system $\{T^{-k} C\}$ is disjoint, and hence $\mu(\bigcup T^{-k} C) = \Sigma \mu(T^{-k} C) \le 1$ by normalization of the measure. But T preserves measure and hence $\Sigma \mu(T^{-k} C) = \sum_n \mu(C)$ and the last series is divergent if $\mu(C) > 0$. Therefore $\mu(C) = 0$.

Property II. ERGODICITY. This property has various formulations, for example: the transformation T is ergodic, if in the theorem of Birkhoff-Khirchin for any function $f \in \mathcal{L}_\mu^1(\mathcal{M})$, the limit function $\hat{f}$ is constant:

$$\lim_{n\to\infty} 1/n \sum_{k=0}^{n} f(T^k x) = \hat{f}(x) = \int_{\mathcal{M}} f(x) d\mu(x) \text{ a. e.}$$

In probability theory this is the form of the most common strengthening of the law of large numbers: The averages converge almost everywhere to the mathematical expectation.

As we will see later, the property of ergodicity implies the indecomposability of the system into nontrivial invariant subsets. More explicitly, the set A is called invariant (invariant mod 0) if $A = T^{-1}A\,(\mu(A \Delta T^{-1}A) = 0)$. We will prove that ergodicity is equivalent to the following assertion: For every invariant mod 0 set A we have that $\mu(A)$ is equal to 0 or 1. We remark that for every invariant mod 0 set there is an invariant set A_1 for which $\mu(A \Delta A_1) = 0$. For this we set first $B = \bigcap_{k=0}^{\infty} T^{-k}A$. It is clear that $\mu(A \Delta B) = 0$ $T^{-1}B \subset B$ and $\mu(B - T^{-1}B) = 0$. The set $A_1 = {}_{k=0}^{\infty} T^{-k}B$ is the one we seek.

Let T be ergodic. We take some invariant mod 0 set A. We find an invariant set A_1 that differs from A in a set of measure 0. For the indicator $\chi_A(x)$ the average $\frac{1}{n}\sum_{k=0}^{n-1} \chi_{A_1}(T^k x)$ is equal to 1 or 0 depending on whether $x \in A_1$ or $x \in A_1$. But it follows from that and ergodicity that if $\mu(A_1) > 0$ holds then the average must equal $\int \chi_{A_1}(x)\,d\mu(x) = \mu(A_1)$, which had to be proved.

Conversely, suppose that $\mu(A) = 0$ or 1 for every set A invariant mod 0 . Then, taking any function $f \in \mathcal{L}^1_\mu(\mathfrak{M})$ and its time average $\hat{f}$, we obtain that for any two numbers a, b, $-\infty < a \leq b < \infty$ the set $E_a^b(\hat{f}) = \{x : a \leq \hat{f} < b\}$ is invariant mod 0. But then $\mu(E_a^b(\hat{f})) = 1$ or 0. This means that $\hat{f}$ = constant almost everywhere.

It follows from the proof that ergodicity is the property of indecomposability of the system into invariant parts.

Property III. Mixing.

The transformation T (automorphism or endomorphism) satisfies the mixing property if for any functions $f, g \in \mathcal{L}^2_\mu(\mathfrak{M})$ the functions $f(T^k x), g(x)$ become statistically independent for large k. Formally, it is sufficient to require

$$\int_{\mathcal{M}} f(T^k x)\cdot g(x)d\mu(x) \xrightarrow[k\to\infty]{} \int_{\mathcal{M}} f(T^k x)\,d\mu(x)\cdot\int_{\mathcal{M}} g(x)d\mu(x) =$$

$$= \int_{\mathcal{M}} f(x)d\mu(x)\cdot\int_{\mathcal{M}} g(x)d\mu(x)\ .$$

For example, if A, B are sets and χ_A, χ_B are their indicators then the independence χ_A and χ_B means that the points that were in the set B initially during that original moment of time distribute themselves evenly over the entire space with the passage of time. That is why, in systems with mixing, it is difficult to distinguish points that were or were not in some set A a sufficiently long time earlier.

Let $\rho(x)$ be a non-negative measurable function for which $\int_{\mathcal{M}} \rho(x)d\mu(x) = 1$. We may introduce a new measure ν_0, absolutely continuous with respect to μ, for which $\frac{d\nu_0(x)}{d\mu(x)} = \rho(x)$. A physicist would call ν a nonequilibrium distribution. We may examine the evolution of the measure ν_0 setting $\nu_n(A) = \nu_0(T^{-n}A)$.

Then in the case of a transformation with mixing

$$\int_{\mathcal{M}} f(x)d\nu_n(dx) = \int f(x)\rho(T^n x)d\mu(x) \xrightarrow[n\to\infty]{} \int f(x)d\mu(x)\ ,$$

which means that the nonequilibrium distribution tends towards equilibrium. We emphasize that ergodicity alone is not sufficient for this.

<u>Property</u> IV. <u>Central Limit Theorem</u> (C. L. T.).

For a given function f we will examine

$$\frac{1}{n}\sum_{k=0}^{n-1} f(T^k x) - \bar{f}, \qquad f = \int_{\mathcal{M}} f(x)d\mu(x)\ .$$

The CLT states that there exists a number $\sigma > 0$ such that

$$\mu\{x: \sigma\sqrt{n}[\frac{1}{n}\sum_{k=0}^{n-1} f(T^k x) - \bar{f}] < a\} \xrightarrow[n\to\infty]{} \frac{1}{\sqrt{2\pi}}\int_{-\infty}^{a} e^{-\frac{\mu^2}{2}} d\mu \tag{2}$$

We will say that the dynamical system satisfies CLT if (2) holds for a sufficiently wide class of functions $f(x)$. Property IV means that the sequence of random variables $f(T^k x)$ behaves like a sequence of independent random variables. Furthermore, CLT shows that the difference $\hat{f}(x) - \frac{1}{n}\sum_{k=0}^{n-1} f(T^x x)$ in the ergodic the theorem of Birkhoff-Khinchin has order $n^{-1/2}$; but it is not possible to write the next term asymptotically: For large n the expression $\sigma\sqrt{n}(\hat{f}(x) - \frac{1}{n}\sum_{k=1}^{n-1} (T^k(x))$ does not converge to any limit, and has only a limit distribution.

Property V. Exponential Decay of Correlation.

Let the function f be such that $\int_{\mathcal{M}} f(x)d\mu(x) = 0$. It is said that exponential decay of correlation holds for the function if there exists a number q, $0 < q < 1$, such that

$$\left|\int_{\mu} f(T^k x)f(x)d\mu(x)\right| \leq C(f)q^k \tag{3}$$

The group $\{T^n\}$ satisfies property (V) if (3) holds for a sufficiently wide family of functions.

The Problem of the Existence of an Invariant Measure.

I. The theory of Bogoliubov-Krilov. We will present only part of this theory. Let $\mathcal{M}$ be a compact topological space and let T be a continuous transformation of $\mathcal{M}$ into itself. Then there always exists at least one measure on $\mathcal{M}$ that is invariant with respect to T.

Proof. Let ν be an arbitrary normalized measure. We set $\nu_k(A) = \nu(T^{-k}A)$ or $T^k\nu = \nu_k$. Furthermore, we introduce the measure $\bar{\nu}_n = \frac{1}{n}\sum_{k=0}^{n-1} \nu_k$. The space of normalized measures on a compact space is weakly compact. This means that a normalized measure μ and a strictly monotone sequence of numbers n_i can be found such that $\lim_{n_i\to\infty} \nu_{n_i} = \mu$. We will prove the invariance of the measure μ, i.e. the equality $\int f(x)d\mu(x) = \int f(Tx)d\mu(x)$ for any bounded measurable function f. By virtue of the compactness of the space $\mathcal{M}$ it is sufficient to verify the last relation only for continuous functions (they form an everywhere dense set in the space of all measurable functions). If f(x) is continuous then

$$\int_{\mathcal{M}} f(x)d\mu(x) = \lim_{n_i\to\infty} \int_{\mathcal{M}} f(x)d\nu_{n_i}(x) = \lim_{n_i\to\infty} \frac{1}{n_i} \sum_{k=0}^{n_i-1} \int_{\mathcal{M}} f(x)d\nu_k(x) =$$

$$= \lim_{n_i\to\infty} \left[\frac{1}{n_i} \sum_{k=1}^{n_i} \int_{\mathcal{M}} f(x)d\nu_k(x) - \frac{1}{n_i} \int_{\mathcal{M}} f(x)d\nu_{n_i}(x) + \frac{1}{n_i} \int_{\mathcal{M}} f(x)d\nu_0(x)\right] =$$

$$= \lim_{n_i\to\infty} \left[\frac{1}{n_i} \sum_{k=1}^{n_i} \int f(T^k x)\ \nu(x)\right] = \lim_{n_i\to\infty} \left[\frac{1}{n_i} \sum_{k=0}^{n_i-1} \int f(Tx)d\nu_k(x)\right] =$$

$$= \int f(Tx)d\mu(x) .$$

Along the way we used

$$\lim_{n_i\to\infty} \frac{1}{n_i} \int_{\mathcal{M}} f(x)d\nu_{n_i}(x) = \lim_{n_i\to\infty} \frac{1}{n_i} \int_{\mathcal{M}} f(x)d\nu_i(x) = 0 .$$

The theorem is proved.

Now we will examine another particular case in which similar considerations apply. Let Q be a compact topological space and suppose that for each $x \in Q$ there is given a function $P(x, \cdot)$ (probability of transition) such that it satisfiesthe following conditions:

1) For every $x \in Q$, $P(x, A)$ is a probability distribution for A.

2) For any $A \in \gamma$, $P(x, A)$ is a measurable function in x.

The collection of functions $P(x, \cdot)$ forms a Markov chain for which Q serves as phase space.

The measure λ is called an invariant measure for the Markov chain with transition probability $P(x, \cdot)$ if for any $A \in \gamma$

$$\lambda(A) = \int_Q P(x, A)\lambda(dx) .$$

The proof of the existence of an invariant measure in the case of a compact Q is very similar to the previous proof. We take an arbitrary measure ν and set $\widetilde{P}\nu(A) = \int_\mu d\nu(x)P(x, A)$. Furthermore, we examine the sequence of averages $\frac{1}{n}\sum_{k=0} \widetilde{P}^k \nu$, from which a weakly convergent subsequence is extracted. Its limit will be the measure sought.

Example.

Suppose that on each integral point of the line there is an automaton which can be in only one of the two states = 0, 1. Thus we have an infinite system of such automata system. We assume that the transition from one state of the system automata into the next is markovian and local. This means that the probability of the state $y_i = x_i(t+1)$ of the ith automaton depends only on the states. $x_{i-1}(t), x_i(t), x_{i+1}(t)$ of the same automaton and of its two closest neighbors and that a function $F(y_i | X_{i-1}, X_i, X_{i+1})$ is determined. The transitions for different automata are independent. Then the expression

$$P(y_{\mu_1}, y_{\mu_2}, \ldots, y_{\mu_k} | x) = \prod_{s=1}^{k} p(y_{\mu_s} | X_{\mu_{s-1}}, X_{\mu_s}, X_{\mu_s+1}) \tag{4}$$

for every $X = \{X_1\}$, $X_1 = 0, 1$ assigns a probability distribution on the space of positions of the automata and generates a Markov chain in such a space. The space of positions of the system automata is the space of infinite binary sequences and is compact in the natural topology. Formula (4) defines a Markov chain and, by what has been proved, there exists an invariant measure; which is far from obvious at first. The following questions become fundamental in such situations:

1. Is the invariant measure unique?
2. How can it be found?

A large number of results was obtained on these problems by I. I. Piatetski-Shapiro and his colleagues: Wasserstein, Leontovich, Stavskaya, Toom, and others.

See, for example, I. I. Pjatecki-Shapiro, O. N. Stawskaya, A. L. Toom, About some homogeneous neuron networks. Automatics and Telemechanics, 1969, N9, 123-128.

Lecture 2.

THE PROBLEM OF THE EXISTENCE OF AN INVARIANT MEASURE

Let $\mathfrak{M}$ be a smooth compact manifold of class C^∞ without boundary, and let T be a diffeomorphism $\mathfrak{M} : \mathfrak{M} \to \mathfrak{M}$ of class C^r $(r \geq 1)$. The theory of Bogoliubov-Krilov yields the existence of at least one invariant measure with respect to T, but that measure may be badly concentrated in separate points, in closed nowhere dense sets of a complicated nature, and having no relation to the smooth structure. Now we will examine certain necessary conditions for the existence of a smooth invariant measure.

We will make a small digression into differential forms. Let $\mathfrak{M}$ be a smooth, closed, and orientable n-dimensional manifold and let T be a smooth transformation on $\mathfrak{M}$. The manifold $\mathfrak{M}$ may be considered to be a two dimensional surface contained in R^3 by those who wish to do so. Through any point x of the manifold $\mathfrak{M}$ it is possible to construct the tangent space $\mathcal{T}_x$. We take it in some ordered basis $e_1, e_2, \ldots, e_n$. Since $\mathfrak{M}$ is orientable, the basis may be oriented positive or negative. Consider the n-dimensional parallelogram $\prod(e_1, e_2, \ldots, e_n)$ spanned by the basis vectors; i.e. the set of all vectors of the form $\Sigma\, \lambda_i e_i$, where $0 \leq \lambda_i \leq 1$, $i = 1, 2, \ldots, n$. We will say that an n-dimensional differential form ω_x is given on $\mathfrak{M}$ if there corresponds a number $\omega_x(\pi)$ to each parallelogram $\pi(e_1, e_2, \ldots, e_n) = \pi$ such that:

1. Under a permutation of the basis vectors ω_x it is multiplied by $(-1)^J$, where $\mathcal{J}$ is the signature of the permutation;

2. $\omega_x(\pi(e_1, \ldots, e_{i-1}, \lambda e_i, e_{i+1}, \ldots, e_n)) = \lambda\omega_x(\pi(e_1, \ldots, e_n))$;

3. $\omega_x(A\pi) = \det A\omega_x(\pi)$, where A is any linear transformation on $\mathcal{T}_x$ (2. is a special case of 3.);

4. ω_x is continuous with respect to x in the obvious sense.

It is possible to integrate differential forms on the manifold $\mathfrak{M}$ by constructing the Riemann sums and passing to the limit. ($\mathfrak{M}$ decomposes into curvilinear parallelepipeds, which are approximated by parallelepipeds in the tangent spaces. Since $\mathfrak{M}$ is orientable, it is possible to choose a decomposition in which every parallelepiped is positively oriented.)

It follows from 3. that, for a fixed x, the value of $\omega_x(\pi)$ is uniquely determined by its value on one parallelepiped π. Therefore, if two non-vanishing n-dimensional differential forms ω and ω_1 are given then there exists a function $\rho(x)$ such that $\omega_1 = \rho\omega$.

If there is a differential form ω of order n and there is a smooth transformation T on the manifold $\mathfrak{M}$ then it is possible to examine the differential form $T^*\omega$ induced by this transformation. It is determined in the following way. Let $x \in \mathfrak{M}$, let $e_1, e_2, \ldots, e_n$ be a basis of the tangent space $\mathcal{T}_x$, and let $y = Tx$.

Suppose, further, that $y_1, y_2, \ldots, y_n$ are local coordinates in a neighborhood of the point y, and that $x_1, x_2, \ldots, x_n$ are local coordinates in a neighborhood of the point x. The transformation T can be represented in the form $y_i = f_i(x_1, \ldots, x_n)$, where f_i is a differentiable function. We will examine the matrix of partial derivatives, called the Jacobian matrix of the transformation T:

$$\frac{\partial f_i}{\partial x_j}\Big|_x : \mathcal{T}_x \to \mathcal{T}_y$$

If π_y is a parallelepiped at the point y then it is possible to examine the parallelepiped at the point x which is related to π_y by the relation

$$\pi_y = \left(\frac{\partial f}{\partial x}\right)\pi_x$$

We will define $T^*\omega$ by the identity $(T^*\omega)_y(\pi_y) = \omega_x(\pi_x)$. Then $T^*\omega$ will again be an n-dimensional differential form. From the remark made above we obtain

$$T^*\omega = \mathcal{J}_{\omega'}(x)\cdot\omega$$

The function $\mathcal{J}_\omega(x)$ is called the Jacobian of the form ω with respect to the transformation T. A smooth measure on $\mathcal{M}$ is given by an n-dimensional differential form: $\mu(A) = \int_A \omega(dx)$.

Our problem is to find a smooth invariant measure, or a corresponding smooth differential form.

The condition of invariance of the form ω consists of $\mathcal{J}_\omega(x) = 1$. If ω_1, $\omega_1 \neq 0$, is an arbitrary form and $\mathcal{J}_{\omega_1}(x)$ is its Jacobian then $\omega = \rho(x)\omega_1$ and, as can be easily verified,

$$\mathcal{J}_\omega = \frac{\rho(x)}{\rho(Tx)}\mathcal{J}_\omega \, .$$

Therefore, ω will be an invariant form if and only if $\mathcal{J}_{\omega_1} = \frac{\rho(Tx)}{\rho(x)}$. We will deduce a series of consequences from this. Let $X^{(0)}$ be a fixed point: $Tx^{(0)} = x^{(0)}$; and let $x_1, \ldots, x_m$ be coordinates in its neighborhood. The transformation T is given in these coordinates by the differentiable functions $f_i(x_1, \ldots, x_n)$. Placing the origin of the coordinates at the point $x^{(0)}$, we will obtain $f_i(0, \ldots, 0) = 0$. Let $\frac{\partial f_i}{\partial x_j}\Big|_0 = a_{ij}$. From this it follows that, in the case of a fixed point, $\mathcal{J}_\omega(x)$ does not depend on ω and $\mathcal{J}_\omega(x) = \det\|a_{ij}\|$. We have obtained a necessary condition for the the existence of an invariant measure: At fixed points $\det\|a_{ij}\| = 1$. Therefore, if $x^{(0)}$ is a contracting point, then $\det\|a_{ij}\| < 1$ and the equality $\mathcal{J}_\omega(x) \equiv 1$ is impossible; i.e. there does not exist an invariant smooth measure; which, however, is clear from obvious considerations: There cannot be a finite, invariant, and smooth measure if the neighborhood of some point is taken strictly inside itself.

Now we will examine an important example which was first investigated by Gauss. The nature of a single valued but not one-to-one, measure preserving transformation is well seen in this example.

Let the interval $[0,1]$ serve as the basic space, and let the transformation T have the form: $Tx = \{\frac{1}{x}\} = \frac{1}{x} - [\frac{1}{x}]$. Let the point $x_0 \in [0,1]$ be fixed. We will find all points x such that $Tx = x_0$. The last relation means that $\frac{1}{x} - [\frac{1}{x}] = x_0$. Let $[\frac{1}{x}] = n$. Correspondingly, we denote the point x by x_n (actually, $\frac{1}{n} \le x_n < \frac{1}{n+1}$). Then $\frac{1}{x_n} - [\frac{1}{x_n}] = \frac{1}{x_n} - n = x_0$, i.e. $x_n = \frac{1}{x_0+n}$, $n = 1,2,\ldots$. In this way, the point x_0 has countably many inverse images because $T[\frac{1}{n+1}, \frac{1}{n}] = [0,1]$, $n = 1,2,\ldots$ and on every such interval there is precisely one point taken to x_0.

We seek the density of the invariant measure in the form $\rho(x_0)\,dx_0$. Gauss showed that $\rho(x_0) = \frac{1}{1+x_0}$. We will verify the invariance of this measure, i.e. that $\rho(x_0)\,dx_0 = \sum_{m=1}^{\infty} \rho(x_n)\,dx_n$. We have

$$\sum_{n=1}^{\infty} \frac{dx_n}{1+x_n} = \sum_{n=1}^{\infty} \frac{dx_n}{1+\frac{1}{x_0+n}} = \sum_{n=1}^{\infty} \frac{x_0+n}{x_0+n+1}\,dx_n .$$

But $\left|\frac{dx_n}{dx_0}\right| = \frac{1}{|f'(x_n)|} = x_n^2$, hence

$$\sum_{n=1}^{\infty} \frac{x_0+n}{x_0+n+1}\,dx_n = \sum_{n=1}^{\infty} \frac{x_0+n}{x_0+n+1}\,x_n^2\,dx_0 = \sum_{n=1}^{\infty} \frac{x_0+n}{(x_0+n+1)(x_0+n)^2}\,dx_0$$

$$= \sum_{n=1}^{\infty} \frac{1}{n+x_0} \cdot \frac{1}{1+n+x_0}\,dx_0 = dx_0 \sum_{n=1}^{\infty} \left(\frac{1}{n+x_0} - \frac{1}{1+n+x_0}\right)$$

$$= \frac{dx_0}{1+x_0}$$

which had to be proved.

The following is a slight variation on the example of Gauss. Let $f(x)$ be given on $[0,1]$, $f(n) = 0$, $f(1) = m$ - a natural number greater than one, and $f'(x) > 1$ for all x. We will define $Tx = \{f(x)\}$. We seek an invariant measure in the form $\rho(x)\,dx$, where the function $\rho(x)$ must satisfy the equation

$$\rho(x_0) = \sum_{x_i : Tx_i = x_0} \frac{\rho(x_i)}{f'(x_i)} .$$

It turns out that this equation is always solvable, and that $\rho(x) \in C^{r-1}$ if $f(x) \in C^r (r \geq 2)$. The last assertion is somewhat more difficult. However, it is easier to prove that $\rho(x)$ is continuous and even satisfies a lipshitz (Hölder) condition, which was done first by A. Renyi (1953). A profound investigation of these transformations was conducted by V. A. Rochlin.

The following example was taken from the recent investigations of Bogayavlenski and Novikov of homogeneous models in the general theory of relativity. Let Δ be an equilateral triangle and let $\mathfrak{M}$ be the circle inscribed in it (see figure). Then $\mathfrak{M}$ is partitioned by the points of tangency into three arcs $\mathfrak{M}_1, \mathfrak{M}_2, \mathfrak{M}_3$ corresponding to the three vertices A_1, A_2, A_3. We will examine the following transformation on $\mathfrak{M}$: A point $x \in \mathfrak{M}_i$ is connected by a straight line segment with A_i, after which that segment is extended to the other side to the next intersection with $\mathfrak{M}$. The point y obtained by this intersection is T_x.

It is easy to see that T is a two-fold covering of $\mathfrak{M}$, the derivative of which is greater than 1 at all points except the tangency points. Bogoyavlenski showed recently that this transformation has a smooth invariant measure of the infinite variation $(\mu(\mathfrak{M}) = \infty)$. We recommend also for a reader interested in similar problems to look at the paper by L. A. Bunimovitsch. On a transformation of the circumference. Matematicheskie Zemetki, 1970, v 8, N2, 205-216.

In the general case, the existence of a smooth invariant measure is a fact that is difficult to establish. The difficulties grow especially in the case of a dynamical system with continuous time.

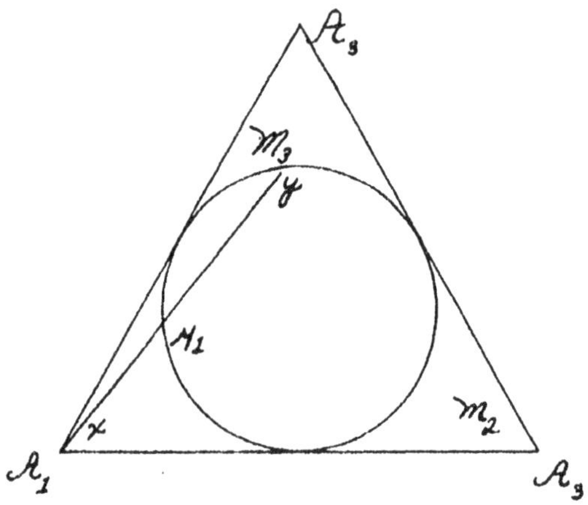

References

1. A. Renyi, Representations for real numbers and their ergodic properties. Acta Math. Acad. Sci. hung. 1957, 8, n3-4, 477-493.

2. V. A. Rohlin, Exact endomorphisms of a Lebesque space. Izvestia Acad. of Sciences, USSR, ser. math. 1961, v25, n. 4, 499-520.

3. P. Billingsley, Ergodic Theory and Information. New York, John Wiley and Sons, 1965.

Lecture 3.

TRANSLATIONS ON COMPACT ABELIAN GROUPS, THEIR APPLICATIONS AND GENERALIZATIONS

In problems of algebraic origin, the existence of an invariant measure follows for algebraic reasons. Now we will examine the simplest examples. Let $\mathfrak{M}$ be a compact abelian group.

Example 1. $S^1 = \{z : z \in \mathbb{C}, |z| = 1\}$.

Example 2. $\mathrm{Tor}^n = S^1 \times S^1 \times \ldots \times S^1$ - the n-dimensional torus.

$$\mathrm{Tor}^n = \{z : z = (z_1, \ldots, z_n) : z_i \in S^1, \ i = 1, 2, \ldots, n\} .$$

Coordinatewise multiplication serves as the group operation in Tor^n.

Ordinary length (Lebesgue measure) serves a measure in the group S^1 , and the measure Tor^n is defined as the direct product measure. This measure is the Haar measure because it is invariant under translations (see below).

We introduce the linear coordinate $x : 0 \leq x \leq 1$ in S^1, where the points $x = 0$ and $x = 1$ are identified. The coordinates on any torus Tor^n are defined analogously.

A translation on a group G is a transformation: $Tg : x \xrightarrow{Tg} gx$. In the case of the torus Tor^n this transformation consists of the translation of the points of the torus by the fixed point $g = (g_1, g_2, \ldots, g_n)$, where $0 \leq g_i \leq 1$, $i = 1, 2, \ldots, n$.

In the coordinates $(x_1, \ldots, x_n)$ the translation is written in the form:

$$x = (x_1, \ldots, x_n) \xrightarrow{Tg} (x_1 + g_1 (\text{mod } 1), \ldots, x_n + g_n (\text{mod } 1)) .$$

In particular for $n = 1$, $x \to x + g \pmod 1$ is, in geometric terms, a rotation of the circle by the angle $2\pi g$. It is clear that the translation preserves the Haar measure on Tor^n. We note an important property of a translation. If we understand the distance between the points x', x'' to be, for example, the length of the shortest arc connecting them $(x', x'' \in S^1)$, or, in case $x', x'' \in \text{Tor}^n$, the sum of such lengths along each coordinate, then the translation Tg is an isometry of the torus Tor^n : $d(Tg^t x', Tg^t x'') = d(x', x'')$ for any integer t.

The set $\{Tg^t x_0, -\infty < t < \infty\}$ is called the trajectory of the point x_0. The following theorem holds.

<u>Theorem.</u> If the trajectory of at least one point x_0 is everywhere dense then the trajectory of every point is everywhere dense.

<u>Proof.</u> We shall write T instead of Tg. Let y be an arbitrary point. We choose $\epsilon > 0$. Then there exists a t_0 such that $\mathcal{L}(T^{t_0} x_0, y) < \frac{\epsilon}{2}$. Since the trajectory of x_0 is everywhere dense, there exists a number N such that the set $\{T^i x_0 : |i| \le N\}$ forms an $\frac{\epsilon}{2}$-net in $\mathcal{M}$.

Since T is an isometry, $d(T^{i+t_0} x_0, T^i y) = d(T^{t_0} x_0, y)$ and so for any point z

$$d(z, T^i y) \le d(z, T^{i+t_0} x_0) + d(T^{i+t_0} x_0, T^i y) .$$

Hence since there exists an i_0, $|i_0 + t_0| \leq N$, for which $d(z, T^{i_0+t_0} x_0) < \frac{\varepsilon}{2}$, we obtain: $d(z, T^i y) < \varepsilon$, i.e. the points $T^i y$, $|t| \leq N + t_0$ form an ε-net. The theorem is proved.

We recall the definition of ergodicity of a transformation given in the first lecture. Let T be a transformation with an invariant measure. It is called ergodic if for almost every point $x \in \mathcal{M}$, and every $f \in \mathcal{L}^1(\mathcal{M}, \mu)$

$$\lim_{n\to\infty} \frac{1}{n} \sum_{k=0}^{n-1} f(T^k x) = \int_{\mathcal{M}} f \, d\mu$$

The expression on the left is called the time average, and the integral $\int_{\mathcal{M}} f \, d\mu$ is called the space average. In the ergodic case the time average coincides almost everywhere with the space average. The function f(x) is called invariant if $f(Tx) = f(x)$ for all $x \in M$. The function f(x) is called invariant mod 0 if

$$f(x) = f(Tx) \text{ a.e.}$$

In the ergodic case every invariant mod 0 function is constant.

<u>Theorem</u> (H. Weyl, von Neumann). The transformation T on the torus Tor^n is ergodic if and only if the equality $\sum_{i=1}^{n} r_i g_i + m = 0$ cannot hold for any integers r_i and m that are not simultaneously equal to zero. (In the one dimensional case S^1 this is equivalent to the irrationality of the number g).

<u>Proof</u>. Let $T = T_g$ be a translation on the torus and let the "irrationality" condition be satisfied. We will prove that f = constant mod 0 if $f \in \mathcal{L}^1(\mathcal{M}, \mu)$

is invariant mod 0 for an arbitrary function $f \in \mathcal{L}^1(\mathfrak{M}, \mu)$. The proof will be broken into three parts.

1. Without loss of generality we may take f to be bounded. If that is not so (i.e. f is not bounded), then for any two a and b we set

$$E_a^b(f) = \{x : a \leq f < b\}$$

The new function $f \cdot \chi_{E_a^b}$ will be bounded and the property of invariance mod 0 will be preserved. If we prove that $f(x)\chi_{E_a^b} =$ constant a.e. then we will obtain the required result by taking the limit $a \to -\infty$, $b \to \infty$.

2. It is possible to expand the bounded function f on the torus in its Fourier series, which converges in the square integral norm:

$$f(x) = \sum_r C_r e^{2\pi i(r, x)}$$

where $r = (r_1, \ldots, r_n)$, $x = (x_1, x_2, \ldots, x_n) \in \mathrm{Tor}^n$ (the r_i are integers) and the sum extends over all such integral n-tuples. The coefficients C_r are determined by the equality: $C_r = \frac{1}{(2\pi)^n} \int_{\mathrm{Tor}^n} e^{-2\pi i(r, x)} f(x)\, dx.$

3, Using the invariance of f mod 0 we have

$$f(Tx) = \sum_r C_r e^{2\pi i(r, x+g)} = \sum_r C_r e^{2\pi i(r, g)} e^{2\pi i(r, x)} = \sum_r C_r e^{2\pi i(r, x)} \pmod{0}.$$

By virtue of the uniqueness of the Fourier coefficients

$$C_r e^{2\pi i(r, g)} = C_r .$$

If $C_r \neq 0$ then $e^{2\pi i(r,g)} = 1$, $(r,g) = m$, where m is an integer. But the last equality is impossible by the hypotheses of the theorem. Hence only one coefficient in the Fourier series, namely C_0, can be different from zero. But this means that $f = C_0 =$ const. mod 0, which had to be proved.

<u>Converse</u>. Let there be a vector with integer coordinates $r = (r_1, \ldots, r_n)$ and an integer m such that $(g, r) + m = 0$, and suppose that some $r_i \neq 0$.

We set $f(x) = e^{2\pi i(x, r)}$. Then

$$f(Tx) = e^{2\pi i(x, r)} \cdot e^{2\pi i(g, r)} = e^{2\pi i(x, r)} e^{2\pi i m} = e^{2\pi i(x, r)} = f(x)$$

The function f is invariant but is not constant. Hence T is not ergodic. The theorem is proved.

The theorem that we have proved is so important that we will present a second proof of it later, but first we will derive a corollary from it.

Let $A = (r_1, r_2)$ be an interval with rational endpoints on the circle S^1. Then, in the case of an ergodic T, for almost every x, $\frac{1}{n} \sum_{k=0}^{n-1} \chi_A(T^k x) \to \ell(A)$, where ℓ is Lebesgue measure. This means that a point falls into every interval with rational endpoints with a frequency that is proportional to the length of the interval. It follows from this and the first theorem of this lecture proved above that the trajectory of every point is everywhere dense. An analogous result holds for the torus Tor^n in case of an ergodic T.

Now we will prove the theorem of Kronecker-Weyl.

Theorem. Let n irrational numbers $g_1, g_2, \ldots, g_n$ be given which are rationally independent together with 1 (i.e. linearly independent over the field of rational numbers). Then for every $\varepsilon > 0$ there is an integer t_0 and integers $K_1, K_2, \ldots, K_n$ such that

$$|t_0 g_i - K_i| < \varepsilon, \quad i = 1, 2, \ldots, n .$$

Proof. The conditions of the theorem are precisely the conditions of the Weyl-von Neumann theorem. We will examine the translation on the torus Tor^n by the vector $g = (g_1, \ldots, g_n)$. This translation is ergodic and hence the trajectory of every point (in particular the point 0) is everywhere dense. But then for every $\varepsilon > 0$ there is an integer t_0 such that $d(T^{t_0} x_0, x_0) <$ where $x_0 = (0, \ldots, 0)$. Since

$$T^t x_0 = (g_1 t_0 (\mathrm{mod}\ 1), \ldots, g_n t_0 (\mathrm{mod}\ 1)) ,$$

the last assertion means precisely that there exists a set of integers $K_1, K_2, \ldots, K_m$ for which the following holds simultaneously:

$$|g_1 t_0 - K_1| < \varepsilon, \ |g_2 t_0 - K_2| < \varepsilon, \ldots, |g_n t_0 - K_n| < \varepsilon .$$

We continue the study of a rotation of the circle: $\mathcal{M}' = S^1$, $Tx = x+g(\mathrm{mod}\ 1)$. This transofrmation, as was shown above, preserves Lebesgue measure and preserves the distance between two points. It follows from the above that the following three conditions are equivalent:

1. The transformation T is ergodic.

2. The number g is irrational.

3. The trajectory of every point is everywhere dense.

Let $\{x_n\}$ be a sequence of points on the unit circle, Δ -an arc on the circle: $\Delta : \{x : \alpha \leq x \leq \beta\}$; we will denote the length of the arc Δ by $\ell(\Delta)$.

<u>Definition.</u> An arbitrary sequence of points $\{x_n\}$ of the unit circle is said to be uniformly distributed on S^1 if for any arc Δ

$$\lim_{n\to\infty} \frac{\nu_n(\Delta)}{n} = \ell(\Delta)$$

where $\nu_n(\Delta)$ denotes the cardinality of the set of numbers $K, 1 \leq K \leq n$, for which $x_K \in \Delta$.

We take the sequence $\{x_n\}$ to be the semitrajectory of some point $x_0 : T^n x_0 = x_n$, $n \geq 0$. If T is ergodic then, by the theorem of Birkhoff-Khinchin, the sequence $\{x_n\}$ is uniformly distributed for almost every initial point x_0. Now we will prove a stronger assertion: the sequence $\{T^n x_0\}$ is uniformly distributed for all x_0 (This assertion is often called Weyl's theorem).

<u>Proof.</u>

I. <u>Assertion.</u> Let $f(x) \in C(S^1)$ be a continuous function on S^1. Then a necessary and sufficient condition for the uniform distribution of the points $\{x_n\}$ is that for every $f \in C(S^1)$

$$\frac{1}{n} \sum_{i=1}^{n} f(x_i) \to \int_{S^1} \cdot f(x)\, d\ell .$$

Indeed, suppose that theconvergence holds for every $f \in C(S^1)$; we will derive from this the uniform distribution of the sequence $\{x_n\}$. Let Δ be an arbitrary arc on S^1, χ_Δ-the indicator function of Δ. Then $\nu_n(\Delta) = \sum_{i=1}^{n} \chi_\Delta(x_i)$, we fix $\epsilon > 0$. Now we choose functions $f^+(x)$ and $f^-(x)$ such that a) they are continuous;

b) $f^-(x) \le \chi_\Delta(x) \le f^+(x)$ for all x;

c) $\int [f^+(x) - f^-(x)]\, dx < \epsilon$ . .

Then $\sum_{i=1}^{n} f^-(x_i) \le \nu_n(\Delta) \le \sum_{i=1}^{n} f^+(x_i)$ or, dividing both parts by n,

$$\frac{1}{n} \sum_{i=1}^{n} f^-(x_i) \le \frac{\nu_n(\Delta)}{n} \le \frac{1}{n} \sum_{i=1}^{n} f^+(x_i) .$$

Fixing the functions f^- and f^+, we let n tend to infinity. Then, according to the hypothesis, the left and right sides tend to $\int f^-\, d\ell$ and $\int f^+\, d\ell$, respectively. This means that

$$\ell(\Delta) - \epsilon \le \underline{\lim}_{n\to\infty} \frac{\nu_n(\Delta)}{\Delta} \le \overline{\lim}_{n\to\infty} \frac{\nu_n(\Delta)}{\Delta} \le \ell(\Delta) + \epsilon .$$

Since ϵ is arbitrary, $\lim\limits_{n\to\infty} \frac{\nu_n(\Delta)}{\Delta}$ exists and equals $\ell(\Delta)$. The proof of necessity is left to the reader.

II. <u>Assertion.</u> Let $f(x)$ be continuous on S^1. Then

$$\frac{1}{n} \sum_{k=1}^{n} f(T^k x_0) = \frac{1}{n} \sum_{k=1}^{n} f(x_0+ky) \xrightarrow[n\to\infty]{} \int f(x)\, d\ell$$

for almost all x_0 $\qquad (*)$

Actually, given a fixed function $f \in C(S^1)$ this relation holds for almost every x_0 by virtue of the ergodicity of the transformation T. Now we will show that it follows from the truth of (*) for at least one initial point x_0 that it is true for every other point $\bar{x}_0$.

Let there be a point x_0 for which $\frac{1}{n}\sum_{k=1}^{n} f(x_0+kg) \xrightarrow[n\to\infty]{} \int f(x)\, d\ell$, and let $\bar{x}_0$ be an arbitrary point on the circle. We choose $\varepsilon > 0$. Since the trajectory of the point x_0 is everywhere dense, there exists an integer $\bar{k}$ such that $d(T^{\bar{k}}x_0, \bar{x}_0) < \delta$, where δ will be determined later. In that case

$$\left|\frac{1}{n}\sum_{k=1}^{n} f(\bar{x}_0+kg) - \frac{1}{n+\bar{k}}\sum_{k=1}^{n+\bar{k}} f(x_0+kg)\right| \le \left|\frac{1}{n}\sum_{k=1}^{n}[f(\bar{x}_0+kg) - f(x_0+(k+\bar{k})g)]\right|$$

$$+ \frac{\bar{k}}{n+\bar{k}} \max_{x\in S^1} |f(x)| + \frac{\bar{k}}{n(n+\bar{k})} \sum_{k=\bar{k}}^{n+\bar{k}} |f(x_0+kg)| \,.$$

The first term can be made less than $\frac{\varepsilon}{2}$ by the choice of δ, and the sum of the other terms on the right side will be small for sufficiently large n. Therefore $\frac{1}{n}\sum_{k=1}^{n} f(\bar{x}_0+kg)$ also converges to $\int f(x)\, d\ell(x)$.

The fundamental point in this proof is the utilization of the fact that if the distance between two points is small at some moment in time then it will remain small at all future moments in time. The uniform distribution of the sequence $\{T^n x_0\}$ for all n follows from the theorem of Birkhoff-Khinchin as well as from the assertions I and II. From the theorem proved we obtain the important

<u>Corollary</u>. The rotation of the circle through an irrational angle has only one invariant Borel measure.

Proof. Let μ be an invariant Borel measure (i.e. a measure defined on Borel sets). The measure μ is uniquely determined by the values of the integrals $\int f(x)\, d\mu(x)$ of continuous functions.

By the ergodic theorem of Birkhoff-Khinchin

$$\lim_{n\to\infty} \frac{1}{n} \sum_{k=0}^{n-1} f(x+kg) = \hat{f}(x) \qquad \mu\text{ - a.e.}$$

By Weyl's theorem we obtain for all x

$$\lim_{n\to\infty} \frac{1}{n} \sum_{k=0}^{n-1} f(x+kg) = \int f\, d\ell .$$

Comparing these two theorems we obtain that

$$\hat{f}(x) = \text{const} = \int f\, d\ell .$$

And so, $\int f\, d\mu = \int f\, d\ell$ for any continuous function $f \in C(S^1)$. Consequently $\mu = \ell$, which was required.

Very recently it was proved by the American mathematician M. Keane that there exists a continuum of continuous pairwise singular quasi-invariant measures for the rotation of the circle through an irrational angle (a measure μ is called quasi-invariant with respect to T if the measure T_μ is equivalent to μ). This result is related to certain problems in the theory of group representations.

References

1. The full metric theory of shifts on compact abelian group was constructed by von Neumann.

See P. Halmos Lectures on Ergodic Theory.

2. V. A. Rohlin, Selected topics of metric theory of dynamical systems. Uspehi math. nayk. 1949, v. 30, N2, 57-126.

3. K. Jacobs, Lecture Notes on Ergodic Theory, vol. 1, 2. Univ. of Aarhus, Denmark, 1962-1963.

Lecture 4

CERTAIN APPLICATIONS OF ERGODIC THEORY TO THE THEORY OF NUMBERS

One of the interesting and important problems of the theory of numbers is the problem of the distribution of the fractional parts of various functions, in particular of polynomials.

Let $f(t)$ be a polynomial of degree $r \geq 1$ with real coefficients, $f(t) = a_0 t^r + a_1 t^{r-1} + \ldots + a_r$. We write $y_n = \{f(n)\}$, $n = 1, 2, 3, \ldots$, where $\{\cdot\}$ is the symbol for the fractional part. It is said that uniform distribution of fractional parts holds for the polynomial $f(t)$ if the sequence $\{y_n\}$ is uniformly distributed on the interval $[0,1]$.

Let $f(t) = a_0 t + a_1$ be a polynomial of first degree.

We have $f(n) = a_0 n + a_1$, $y_n = \{a_0 n + a_1\} = T^n a_1$, where $Tx = x + a_0$ (mod 1). It follows from the proof of an earlier theorem that the sequence y_n is uniformly distributed if a_0 is irrational. The general theorem of H. Weyl is valid for polynomials of arbitrary degree.

Theorem (H. Weyl). If among the coefficients $a_0, a_1, \ldots, a_{n-1}$ there is at least one irrational number then the sequence $\{y_n\}$ is uniformly distributed.

We will not prove this theorem, but for $r = 2$ we will clarify how the matter can be reduced to ergodic theory. And so, let

$$f(t) = a_0 t^2 + a_1 t + a_2 .$$

We will take the two dimensional torus $\mathrm{Tor}^2 =$ [illegible], i.e. the unit square with pairwise identified sides. We take the usual Lebesque measure as the measure on Tor^2.

We define the transformation $T(x, y)$ on the torus Tor^2, called a skew translation or a skew rotation, setting $T(x, y) = (x+\alpha \pmod 1), y+2x+\alpha \pmod 1))$, where the number α will be determined later. The skew translation transforms the circle $x = \text{const}$ into the circle $x = \text{const} + \alpha \pmod 1$ and translates it parallel to itself through the angle $2x + \alpha$, depending on x. T preserves Lebesque measure since it is linear, and the matrix of its first derivatives is triangular and its determinant equals 1. We will prove the following formula by induction:

$$T^n(x, y) = (x+n\alpha \pmod 1), \alpha n^2+2nx+y \pmod 1)).$$

If $n = 1$ then it is true by virtue of the definition of the transformation T. Suppose that it is true for some n. We will verify its truth for $n + 1$. We have

$$T^{n+1}(x, y) = T(T^n(x, y)) = T(x+n\alpha, \alpha n^2+2nx+y) = (x+(n+1)\alpha, \alpha n^2+2nx+y+2(x+n\alpha)+\alpha)$$

$$= (x+(n+1)\alpha, y+2(n+1)x+\alpha(n+1)^2) .$$

Now we will establish the following fact: If α is irrational then the transformation T is ergodic.

The proof that we will present is similar to the analogous proof for the rotation of the circle, although there is something new in it.

Let $h(x, y)$ be a function on the torus Tor^2 invariant mod 0 with respect to T, i.e. $h(x, y) = h(T(x, y))$ a.e. . As we have seen, the assumption of boundedness of the modulus of the function h does not reduce the generality of our reasoning. Therefore it may be assumed that h is bounded and can be expanded in its Fourier series: $h = \Sigma e_{n_1 n_2} e^{2\pi i(n_1 x_1 + n_2 x_2)}$, converging in the square integral norm. Under this

$$\int |h(x, y)|^2 \, dx \, dy = \Sigma |e_{n_1 n_2}|^2 < \infty .$$

It follows from the last inequality that among the Fourier coefficients of the function h there cannot be an infinite number of nonzero coefficients equal in modulus. This is the new part that we will use below in an essential way.

$$h(T(x, y)) = \Sigma c_{n_1 n_2} e^{2\pi i(n_1(x+\alpha)+n_2(y+2x+\alpha))} = \Sigma e_{n_1 n_2} e^{2\pi i(n_1+n_2)} \cdot e^{2\pi i((n_1+2n_2)x+n_2 y)}$$

which be virtue of invariance mod 0 coincides with $\Sigma e_{n_1 n_2} e^{2\pi i(n_1 x+n_2 y)}$ almost everywhere. We will show that the equation

$$\Sigma e_{n_1 n_2} e^{2\pi i(n_1+n_2)\alpha} e^{2\pi i((n_1+2n_2)x+n_2 y)} = \Sigma c_{n_1 n_2} e^{2\pi i(n_1 x+n_2 y)}$$

does not have a solution under the conditions $\Sigma |e_{n_1 n_2}|^2 < \infty$. We make a change of notation: $n_1 + 2n_2 = m_1$, $n_2 = m_2$. Then by the uniqueness theorem for Fourier coefficients we have

$$e_{m_1-2m_2, m_2} e^{2\pi i(m_1-m_2)a} = e_{m_1, m_2}$$

From this $|e_{m_1-2m_2, m_2}| = |e_{m_1, m_2}|$.

It is clear from the last equality that if $m_2 \neq 0$ can be found for which at least one coefficient e_{m_1, m_2} does not equal zero then there will be infinitely many nonzero coefficients in the Fourier series that are equal in modulus, which is impossible. Consequently, $e_{m_1, m_2} = 0$ if $m_2 \neq 0$.

And so, only the coefficients $e_{m_1, 0}$ can be different from zero.

This means that the function $h(x, y)$ depends only on x:

$$h(x, y) = h(x), \ h(T(x, y)) = h(x+a) .$$

By hypothesis the function h is invariant mod 0. If a is irrational then the rotation through the angle $2a\pi$ is ergodic. That means $h = \text{const}$ almost everywhere, from which the ergodicity of the skew translation follows.

We return to the Weyl theorem on the distribution of the fractional parts for the case of a polynomial of second degree. We have

$$T^n(x, y) = (x+na, n^2a+2nx+y), \quad f(t) = a_0t^2 + a_1t + a_2 .$$

We set $a_0 = a$, $a_1 = 2x$, $a_2 = y$. Then the second coordinate of the point $T^n(x, y)$ is equal to the fractional part of the value $f(t)$ at the point n: $\{an^2 + 2nx + y\}$. For irrational a the transformation T is ergodic, which follows from the assertion just proved. It can be shown (we will not do this now) that the trajectory of every point (x, y) is uniformly distributed on the torus, and that T always has an invariant measure - and a unique one at that. The proofs of this theorem that are known to me are based on number theoretic considerations. Another proof can be found in the paper by M. Furstenberg cited below. The equality

$$\lim_{n\to\infty} \frac{1}{n} \sum_{k=1}^{n} f(T^k(x, y)) = \int f \, dx \, dy ,$$

valid for all points (x, y) and every continuous function f, follows from the uniqueness of the invariant measure. Let $f(x, y) = f(y)$ be a continuous periodic function on S^1. Then $f(T^k(x, y)) = f(k^2\alpha+2kx+y) = f(y_k)$ and hence for irrational α

$$\lim_{n\to\infty} \frac{1}{n} \sum_{k=1}^{n} f(y_k) = \int f(y) \, dy \quad .$$

With that we have come to the particular case of Weyl's theorem. The case where α_0 is rational and α_1 is irrational is even easier to examine.

The problem of the distribution of the fractional parts of polynomials is related to the calculation of the number of solutions to diophantine equations in the algebraic theory of numbers.

The interested reader can find further information in H. Furstenberg, Strict ergodicity and transformation of the torus. Amer. J. of Math., 1961, v. 83, NY, 573-601, and A. G. Postnikov, Ergodic problems of the theory of congruences and Diophant approximations. Proceedings of Steklov Institute, v. 82, 1966.

Lecture 5

A SECOND PROOF OF THE ERGODICITY OF THE ROTATION OF A CIRCLE AND PERMUTATIONS

We will give a new proof of the ergodicity of the rotation of a circle through an irrational angle without using the expansion of a function into its Fourier series.

Let, as usual, $Tx = x + g \pmod 1$ and let g be an irrational number. Taking an arbitrary point x_0 we will examine its trajectory $\{T^n x_0\}$, $n = 0, \pm 1, \pm 2, \ldots$. We will prove that the closure $\overline{\{T^n x_0\}} = S^1$. Indeed, without loss of generality we may consider that $x_0 = e$ - the identity of the group S^1. Then $\{T^n x_0\}$ is a subgroup of the group S^1, and $\overline{\{T^n x_0\}}$ is a closed subgroup of S^1. It is known that every closed subgroup of S^1 is either the group of nth roots of unity or the whole group (see, for instance, the book "Continuous Groups" by L. S. Pontryagin). In the first case g is clearly rational. Hence $\overline{\{T^n x_0\}} = S^1$ in our case. But then the trajectory of every point is everywhere dense on S^1.

Now we will indicate an outline of the proof of the ergodicity of T, leaving it to the reader to construct the proof in detail.

Let A and B be invariant sets with respect to T such that $0 < \mu(A) < 1$, $0 < \mu(B) < 1$, $A \cap B = \phi$. Let x_0 be a point of density of the set A. This means that

$$\lim_{\varepsilon \to 0} \frac{\ell((x_0-\varepsilon, x_0+\varepsilon) \cap A)}{2\varepsilon} = 1,$$

where $(x_0-\varepsilon, x_0+\varepsilon)$ is an arc with center at the point x_0. We recall here the theorem that states that almost all points of an arbitrary set with positive measure on the circle are points of density of that set (see, for example, I. P. Nathanson's "Introduction to the theory of functions of a real variable"), from which the existence of the point x_0 follows. Let y_0 be a point of density of the set B. A small neighborhood of the point y_0 consists basically of points of the set B. Now we translate a neighborhood of x_0 with the help of a power of T. Since $\{T^n x_0\}$ is everywhere dense in S^1, for some k the point $T^k x_0$ will fall into an arbitrary neighborhood of the point y_0. But the point $T^k x_0$ is a point of density of the set A because T is a rotation:

$$\frac{\ell(T^k(x_0-\varepsilon, x_0+\varepsilon)\cap A)}{2\varepsilon} = \frac{\ell(T^k((x_0-\varepsilon, x_0+\varepsilon)\cap A))}{2\varepsilon} = \frac{\ell((x_0-\varepsilon, x_0+\varepsilon)\cap A)}{2\varepsilon}$$

by virtue of the hypothesized invariance of that set. But that contradicts the fact that $A \cap B = \phi$.

We turn now to the examination of a new important class of transformations

Permutations. The interval $[0,1]$ with the usual Lebesque measure serves as phase space. Let $[0,1]$ be partitioned into intervals $\Delta_1, \Delta_2, \ldots, \Delta_p$ of arbitrary length whose union is $[0,1]$ and which are pairwise disjoint except for their endpoints. Permutation is a transformation on $[0,1]$ consisting of a rearrangement of these intervals in a different order. We will clarify the importance of the study of permutations.

Many natural measure spaces are isomorphic as measure spaces to the interval $[0,1]$ with the usual Lebesgue measure. The measure preserving transformations T of the interval are given by certain measurable functions $f:[0,1]\to[0,1]$. Such functions can be approximated by a piecewise linear function yielding transformations of the permutation type (here an inclination of -1 corresponds to the translation and symmetric reflection of the interval).

In this way, permutations can be used to approximate arbitrary measure preserving transformations on the interval. In addition, we will see, the permutations themselves arise in certain interesting examples of dynamical systems.

Let T be a permutation of the intervals $\Delta_1, \Delta_2, \ldots, \Delta_p$. We will examine the transformation T^2. It is clear that it consists of a permutation of the intervals $\Delta_{ij} = \Delta_i \cap T\Delta_j$. Analogously, we can examine the intersection $\Delta_{i_0} \cap T\Delta_{i_1} \cap \ldots \cap T^{n-1}\Delta_{i_{n-1}} \cap T^n\Delta_{i_n}$ and determine T^n as a permutation of intervals of this type.

<u>Definition.</u> The permutation T is called transitive if

$$\lim_{n\to\infty} \max_{i_0, i_1, \ldots, i_n} \ell(\Delta_{i_0} \cap T\Delta_{i_1} \cap \ldots \cap T^n\Delta_{i_n}) = 0 .$$

Before formulating and proving a theorem on permutations due to Oseledetz, we turn to certain facts related to unitary matrices and their generalizations - unitary operators on Hilbert space.

Let H be a finite dimensional complex Hilbert space. We will say that a unitary operator U acts on H if U is an invertible linear transformation of the space H which preserves the scaler product: for any h_1 and h_2 in H $(Uh_1, Uh_2) = (h_1, h_2)$.

A unitary operator U has eigenvalues and a complete set of orthogonal eigenfunctions. All eigenvalues of a unitary operator have absolute value one. The situation in which several eigenfunctions correspond to the same eigenvalue is possible: $Ue_1 = ze_1, Ue_2 = ze_2, \ldots, Ue_k = ze_k, e_i \perp e_j$.

The set of such eigenfunctions spans a vector space. The dimension of this vector space is called the multiplicity of the eigenvalue z. The vector spaces H_s corresponding to distinct eigenvalues z_s are pairwise orthogonal and $H = \bigoplus_s H_s$.

Let h be an arbitrary vector in H. We will examine the vector space spanned by the vectors $\ldots U^{-2}h, U^{-1}h, h, \ldots, U^n h, \ldots,$ which we denote $H(h)$. This vector space is finite dimensional since it is contained in H. In addition, $H(h)$ is invariant in the sense that $H(h) = UH(h) = U^{-1}H(h)$.

It turns out that $H(h)$ can be obtained in a different way. Let h_i be the projection of h on the eigenspace H_i corresponding to the eigenvalue z_i of the unitary operator U. Then $H(h)$ is the linear hull of the vectors h_i. It is best to verify this fact directly. From this we obtain the important

Corollary. Let there be given vectors $h^{(1)}, h^{(2)}, \ldots, h^{(k)}$ and let $H = \sum_{\ell=1}^{k} H(h^{(\ell)})$. Here the subspaces $H^{(\ell)} = H(h^{(\ell)})$ need not be orthogonal. Then $\dim H_i \leq K$ for every i (in other words, the multiplicity of every eigenvalue does not exceed K).

Proof. We denote by $h_i^{(\ell)}$ the projection of the vector $h^{(\ell)}$ on to the subspace H_i. By hypothesis the system $\{h_i^{(\ell)}, \ell = 1, 2, \dots, k\}$ generates the subspace H_i. But that means that $\dim H_i \leq k$.

We will generalize our reasoning to the infinite dimensional case. Let H be a separable, infinite dimensional Hilbert space and let U be a unitary operator acting on H.

In this case U may not have eigenvalues, and moreover, if they exist then their multiplicity is at most countable.

By analogy with the finite dimensional construction we denote by H(h) the closure of the linear hull of vectors of the form $U^k h$, $k \in \mathbb{Z}$. Then H(h) is the smallest subspace containing the vector h which is invariant with respect to U and U^{-1}.

Let $\{Z_n\}$ be the eigenvalues of the operator U, and let $\{H_n\}$ be the corresponding subspaces that are invariant under U. We write $h_n = P_{H_n}(h)$, where P is the projection operator.

Lemma. Suppose that there exist finitely many vectors $h^{(1)}, \dots, h^{(k)}$ such that $\sum_{\ell=1}^{k} H(h^{(\ell)}) = H$. Then for all ℓ $\dim H_\ell \leq k$.

The proof repeats exactly the proof for the finite dimensional case. We limit ourselves to this information about unitary operators and turn to the proof of the following theorem.

Theorem (V. Oseledetz). A transitive permutation of the p intervals $\Delta_1, \Delta_2, \dots, \Delta_p$ always has no more than p invariant sets with positive measure such that T is indecomposable (i.e. ergodic) on each of them.

Proof. The role of H will be played by the space $\mathcal{L}^2([0,1])$-the space of square integrable functions on the interval $[0,1]$ with Lebesgue measure. The operator U on it is defined, as usual, by the identity

$$(Uh)(x) = h(Tx) .$$

The invertibility of U follows from the invertibility of T, and U is a unitary operator because T preserves measure. We will examine the following functions:

$$h^{(1)} = \chi_{\Delta_1}, h^{(2)} = \chi_{\Delta_2}, \ldots, h^{(p)} = \chi_{\Delta_p} .$$

Now we will show that $\sum_{i=1} H(h^{(i)}) = H$ (*). Then it will follow from the lemma that the multiplicity of every eigenvalue of the operator U does not exceed p. We will derive the required assertion from this. Indeed, let there be s invariant sets $A_1, A_2, \ldots, A_s$ such that $A \cap A_j = \phi$ if $i \neq j$ and $m(A_i) > 0$, $i = 1, 2, \ldots, s$. Since the A_i are pairwise disjoint, the functions $\chi_{A_i}(x)$ are pairwise orthogonal. In addition, $\chi_{A_i}(x)$ is invariant with respect to T because A_i is invariant mod 0 with respect to T. For all i $U\chi_{A_i} = \chi_{A_i}$. Hence the dimension of the invariant subspace H_1 corresponding to the eigenvalue $1 = z$ is greater than or equal to s. But by the lemma and from the identity (*) we have: $s \leq \dim H_1 \leq p$.

Now we will turn to the proof of the identity (*). We will examine the functions $U^{-1}\chi_{\Delta_1} = \chi_{T\Delta_1}$, $U^{-1}\chi_{\Delta_2} = \chi_{T\Delta_2}, \ldots, U^{-1}\chi_{\Delta_p} = \chi_{T\Delta_p}$. We will prove that all possible functions $\chi_{\Delta_{i_0} \cap T\Delta_{i_1}}$ are linear combinations of the functions χ_{Δ_i} and $U^{-1}\chi_{\Delta_j}$. Indeed, suppose that the last interval Δ_p is covered by some number of intervals $T\Delta_{i_1}, T\Delta_{i_2}, \ldots, T\Delta_{i_k}$ and that $\Delta_p \supset T\Delta_{i_1} \cup T\Delta_{i_2} \cup \ldots \cup T\Delta_{i_{k-1}}$, $\Delta_p \subset T\Delta_{i_1} \cup T\Delta_{i_2} \cup \ldots \cup T\Delta_{i_k}$. Then $\Delta_p \cap T\Delta_{i_s} = T\Delta_{i_s}$ if $1 \leq s \leq k-1$. Furthermore, $\chi_{\Delta_p \cap T\Delta_{i_k}} = \chi_{\Delta_p} - \sum_{s=1}^{k-1} \chi_{T\Delta_s}$. If $T\Delta_{i_k} \subset \Delta_p \cup \Delta_{p-1}$ then $\chi_{\Delta_{p-1} \cap T\Delta_{i_k}} = \chi_{T\Delta_{i_k}} - \chi_{\Delta_p \cap T\Delta_{i_k}}$ and so forth.

Carrying such arguments further we would obtain that $\chi_{\Delta_i \cap T\Delta_j}$ is expressible in the form of a linear combination of functions $\chi_{\Delta_i}, \chi_{T\Delta_j}$. In the same way we would show that $\chi_{\Delta_{i_1} \cap T\Delta_{i_2} \cap \ldots \cap T^{k-1}\Delta_{i_{k-1}} \cap T^k\Delta_{i_k}}$ is expressible as a linear combination of functions $\chi_{\Delta_{i_0}}, \chi_{T\Delta_{i_1}}, \ldots, \chi_{T^k\Delta_{i_k}}$.

But then the function $\chi_{\Delta_{i_1} \cap \ldots \cap T^k\Delta_{i_k}}$ belongs to the vector space $\overline{\sum_{i=1}^{p} H(\chi_{\Delta_i})}$. Since the permutation T is transitive, linear combinations of functions $\chi_{\Delta_{i_1} \cap \ldots \cap T^k\Delta_{i_k}}$ are everywhere dense in $\mathcal{L}^2([0,1])$. (See Halmos p. 171-4.)

Consequently: $\sum_{i=1}^{\ell} H(\chi_{\Delta_i}) = \mathcal{L}^2([0,1])$. The theorem is proved.

Here are some references related to the material of this lecture:

1. Chacon, Approximation of Transformations with continuous spectrum. Pacific J. of Math., 1969, v. 31, N2, 293-303.

2. V. I. Oseledetz, On spectrum of ergodic automorphisms, Doklady, 1966, v. 168, N5, 1009-1011.

3. A. B. Katok and A. M. Stepin, On approximations in ergodic theory, Uspeki, Mat. Nauk. 1967, v. 22, N5, 81-106.

4. P. Halmos, Measure Theory.

Lecture 6

DYNAMICAL SYSTEMS WITH CONTINUOUS TIME

Let $\mathcal{M}$ be a measurable space with the natural distinguished σ-algebra γ of measurable subsets. Let there be defined for every t, $-\infty < t < \infty$, a transformation S_t of the space $\mathcal{M}$ into itself which satisfies

1) For all t_1, t_2 $S_{t_1} \circ S_{t_2} = S_{t_1+t_2}$;

2) For every function $f(x)$ that is measurable with respect to γ the function $g(x,t) = f(S_t x)$ is measurable on the direct product space $\mathbb{R}^1 \times \mathcal{M}$.

In this case the continuous one-parameter group $\{S_t\}$ is called a dynamical system with continuous time or, sometimes, a flow.

The measure μ is called an invariant measure with respect to the group $\{S_t\}$ if for every $A \in \gamma$ $\mu(S_t A) = \mu(A)$ for all t.

We will examine only normalized invariant measures for which $\mu(\mathcal{M}) = 1$.

The theorem of Birkhoff-Khinchin for measure preserving flows reads as follows: For every function $f \in \mathcal{L}^1_\mu(\mathcal{M})$ the limit below exists with probability one:

$$\lim_{T\to\infty} \frac{1}{T} \int_0^T f(S_t x)\, dt = \lim_{T\to\infty} \frac{1}{T} \int_0^T f(S_{-t} x)\, dt = \hat{f}(x) ,$$

and $\int_{\mathcal{M}} f \, d\mu = \int \hat{f} \, d\mu$.

The flow $\{S_t\}$ is called ergodic if $\hat{f} = \text{const} = \int f \, d\mu \pmod 0$ for every function $f \in \mathcal{L}^1(\mathcal{M}, \mu)$.

In the ergodic case it follows from the Birkhoff-Khinchin ergodic theorem that almost every trajectory of an ergodic flow spends an amount of time proportional to $\mu(A)$ in the measurable set A which is proportional to $\mu(A)$. The theory of Bogoliubov-Krilov carries over to the continuous case without any changes.

We will examine an important example. Let $\mathcal{M} = \mathrm{Tor}^n = \underbrace{S^1 \times S^1 \times \ldots \times S}_{n \text{ times}}$ i.e. the n-dimensional torus, $z = (z_1, \ldots, z_n) \in \mathrm{Tor}^n$. A point of the torus Tor^n can be given in multiplicative notation as a choice of complex numbers z_k, $|z_k| = 1$, $1 \leq k \leq n$, and can also be given additively: $z_k = e^{2\pi i \varphi_k}$ and $z = (\varphi_1, \ldots, \varphi_n)$ where $\varphi_i^1 = \varphi_i + m_i \leq$ if m_i is integral and φ_i denote one and the same point. Let the sequence of real numbers $\lambda_1, \lambda_2, \ldots, \lambda_n$ be given.

We take Haar measure $d\mu = \prod_{i=1}^{n} d\varphi_i$ for the measure on $\mathcal{M} = \mathrm{Tor}^n$. We define the flow on Tor^n with the relation $S_t z = (e^{2\pi i z_1 t} \cdot z_1, \ldots, e^{2\pi i \lambda_n t} z_n)$ or $S_t(z) = (\varphi_1 + \lambda_1 t (\mathrm{mod}\ 1), \ldots, \varphi_n + \lambda_n t (\mathrm{mod}\ 1))$.

It is clear that the flow $\{S_t\}$ preserves Haar measure.

The set of vectors $a(t) = (e^{2\pi i \lambda_1 t}, \ldots, e^{2\pi i \lambda_n t})$, $-\infty < t < \infty$ is the trajectory of zero of the torus Tor^n looked at as a group. It forms a one-parameter subgroup of the torus Tor^n, and hence $\{S_t\}$ is a group of translations on the torus Tor^n in the direction of the subgroup $\{a(t)\}$. The flow $\{S_t\}$ is often called conditionally periodic, and λ_i-its frequencies.

<u>Theorem.</u> A conditionally periodic flow is ergodic if and only if the numbers $\lambda_1, \lambda_2, \ldots, \lambda_n$ are linearly independent over the field of rational numbers.

Proof. First we remark that in the case of a flow on the torus, as in the discrete case, the trajectory of all points are simultaneously everywhere dense on the torus. Let $f(z) = f(S_t z) \bmod 0$, where the function f can be considered bounded. It may be expanded in a Fourier series that converges in the square integral norm.

In the additive notation the function on the torus can be written as a function of the variables $\varphi_1, \ldots, \varphi_n$ that is periodic in each variable with period 1, and can be expanded in a complex Fourier series,

$$f(z) = f(\varphi_1, \ldots, \varphi_n) = \sum_{m_1, \ldots, m_n} e^{2\pi i(m_1\varphi_1 + \ldots + m_n\varphi_n)} e_{m_1, \ldots, em_n},$$

$$f(s_t z) = f(\varphi_1 + \lambda_1 t, \ldots, \varphi_n + \lambda_n t) = \sum_{\vec{m}} e_{\vec{m}} e^{2\pi i(\vec{m}, \lambda)t} \cdot e^{2\pi i(\vec{m}, \varphi)} = f(z) .$$

If $(\vec{m}, \lambda) \neq 0$ for any $\vec{m} \neq 0$ then the last equality cannot take place (if t is arbitrary), and all the arguments of the corresponding theorem for the discrete case can be repreated.

In the other direction the proof is also conducted through the same path taken in the discrete case.

In connection with the last theorem, we draw attention to the fact that the ergodicity of individual transformations S_t that enter into the flow does not follow from the ergodicity of the flow $\{S_t\}$.

Indeed, let $n = 1$ and let $S_t x$ be the motion of the point x on the circle with constant velocity $v = 1$ (we recall that $\mu(S^1) = 1$). It is clear that this flow is ergodic. On the other hand, $S_n = \mathrm{Id}$, where Id is the identity transformation.

Now we turn to the study of one of the applications of flows on the torus which arises from problems of celestial mechanics and which has stimulated the development of the theory of almost periodic functions.

The theorem of Lagrange on mean motion.

Let there be given n complex numbers $a_1, a_2, \ldots, a_n$ (n vectors in the plane). We will examine the curve on the complex plane of the variable

$$z(t) = a_1 e^{2\pi i\lambda_1 t} + a_2 e^{2\pi i\lambda_2 t} + \ldots + a_n e^{2\pi i\lambda_n t} .$$

The geometric meaning of the function $z(t)$ is such: Imagine that there is a vector a_1 in the plane, that the vector a_2 is attached thus to the end of a_1, and so forth. Let a_1 turn around its origin with the angular velocity λ_1, let a_2 turn at the same time around its origin (i.e. the point a_1) with angular velocity λ_2, and so forth. $z(t)$ is the trajectory of the end of the vector a_n.

We suppose that $z(t)$ does not become zero for any t. Then we may write

$$z(t) = r(t)e^{2\pi i\varphi(t)} ,$$

where $\varphi(t)$ is a continuous function of t.

The problem of Lagrange consists of the following: Does there exist a limit $\omega = \lim_{T\to\infty} \frac{1}{T}\varphi(T)$ and what does it equal; in other words, with what mean angular velocity does the end of the vector a_n turn around the origin of the vector a_1?

Lagrange himself obtained the answer for the case of two vectors. If $|a_2| + |a_3| + \ldots + |a_n| < |a_1|$ then it is clear that $\omega = \lambda_1$. We will examine the general case and we will show how this problem leads to a problem in ergodic theory. We will not, however, give a completely rigorous foundation. It is clear from the expression for $z(t)$ that $\varphi(t) = \mathrm{Re}(\frac{1}{2\pi i} \ln z(t))$, from which

$$\frac{d\varphi}{dt} = \mathrm{Re}\left(\frac{1}{2\pi i}\frac{z'(t)}{z(t)}\right) = \mathrm{Re}\frac{\sum_{s=1}^{n} \lambda_s a_s e^{2\pi i\lambda_s t}}{\sum_{s=1}^{n} a_s w^{2\pi i\lambda_s t}} = \mathrm{Re}\frac{\Sigma |a_s| \lambda_s e^{2\pi i(\lambda_s t + \bar{\varphi}_s)}}{\Sigma |a_s| e^{2\pi i(\lambda_s t + \bar{\varphi}_s)}},$$

where $\bar{\varphi} = (\bar{\varphi}_1, \ldots, \bar{\varphi}_n)$ is the original point and

$$z(t) = |a_1| e^{2\pi i(\bar{\varphi}_1 + \lambda_1 t)} + \ldots + |a_n| e^{2\pi i(\bar{\varphi}_n + \lambda_n t)}.$$

Now we take the torus Tor^n and we examine the conditionally periodic flow on it that is determined by the vector with components $\lambda_1, \ldots, \lambda_n$.

We suppose that the numbers $\lambda_1, \ldots, \lambda_n$ are linearly independent over the field of rational numbers. Then our flow is ergodic. On the torus Tor^n we examine the function:

$$f(\varphi) = f(\varphi_1, \varphi_2, \ldots, \varphi_n) = \mathrm{Re}\frac{\sum_{s=1}^{n} \lambda_s |a_s| e^{2\pi i\varphi_s}}{\sum_{s=1}^{n} |a_s| e^{2\pi i\varphi_s}} \qquad (1)$$

It is clear that the following equality holds

$$\frac{d\varphi}{dt} = f(\varphi)\big|_{\bar{\varphi} + \bar{\lambda} t} = f(\varphi_t) ,$$

where

$$\bar{\varphi} = (\varphi_1, \varphi_2, \ldots, \varphi_n), \quad \bar{\lambda} = (\lambda_1, \ldots, \lambda_n) .$$

Hence $\varphi(t_1) - \varphi(t_2) = \int_{t_1}^{t_2} f(\varphi_n)\, dn$.

We must find $\lim_{t\to\infty} \frac{\varphi(t)}{t} = \lim_{t\to\infty} \frac{1}{t} \int_0^t f(\varphi_n(dn$.

If the function $f(\varphi)$ were bounded and continuous then such a limit would exist everywhere and would equal $\int_{\mathrm{Tor}^n} f(\varphi)\, d\mu$ by the ergodic theorem. However, the denominator in the right side of (1) can become zero. The condition for the vanishing of the denominator $\sum_{s=1}^{n} |a_s| e^{2\pi i \varphi_s} = 0$ contains as a matter of fact two equations (for real and imaginary parts) and determines a submanifold of codimension 2 on the torus Tor^n. It follows from this that a typical trajectory does not pass through this special manifold.

Suppose that the ergodic theorem applies. Then

$$\int_{\mathrm{Tor}^n} f(\varphi)\, d\varphi = \mathrm{Re} \int_{\mathrm{Tor}^n} \frac{\Sigma \lambda_s |a_s| e^{2\pi i \varphi_s}}{\Sigma |a_s| e^{2\pi i \varphi_s}}\, d\varphi_1 \ldots d\varphi_n = \Sigma \lambda_s |a_s| \cdot w_s ,$$

$$w_s = \mathrm{Re} \int_{\mathrm{Tor}^n} \frac{e^{2\pi i \varphi_s}}{\sum_{t=1}^{n} |a_t| e^{2\pi i \varphi_t}}\, d\varphi_1 \ldots d\varphi_n .$$

We must compute w_s. To that end, we fix the values of all the variables $\varphi_1, \ldots, \varphi_n$ with the exception of the variable φ_s.

Then:

$$w_s = \int_{\mathrm{Tor}^n} \left[\int \frac{e^{2\pi i \varphi_s} \, d\varphi_s}{B(\varphi_1, \ldots, \varphi_{s-1}, \varphi_{s+1}, \ldots, \varphi_n) + |a_s| e^{2\pi i \varphi_s}} \right] d\varphi_1 \ldots \widehat{d\varphi_s} \ldots d\varphi_n .$$

The expression under the integral sign can be written in the form

$$\frac{1}{|a_s| 2\pi i} \frac{d \ln(B + |a_s| e^{2\pi i \varphi_s})}{d\varphi_s} .$$

The equation for the circle stands under the logarithm sign.

Only two cases are possible

I. The origin of the coordinates lies inside the circle.

II. The circle does not contain the origin of the coordinates.

In case I the integral equals one, and in case II it becomes zero.

The condition that the circle contains the origin within itself corresponds to the inequality $|B| \leq |a_s|$. Hence

$$\int \frac{e^{2\pi i \varphi_s} d\varphi_s}{B + |a_s| e^{2\pi i \varphi_s}} = \frac{1}{2\pi i |a_s|} \cdot P\{|a_s| > |B|\} .$$

This result can be interpreted in the following way. w_s is the fraction of moments in time when the rotation of the vector a_s makes a contribution to the function $\varphi(t)$.

References

1. See S. Sternberg, "Celestial Mechanics," Part I.

Lecture 7

LINEAR HAMILTONAIN SYSTEMS

The aim of today's lecture of parts of subsequent lectures consists in convincing the listener that the dynamical systems on the torus are met in a large number of problems of ergodic theory.

Let $\mathfrak{M}^n$ be an n-dimensional differential manifold of class C^∞. We denote by $\mathfrak{T}_x$ the tangent space to the manifold $\mathfrak{M}^n$ at the point x. By $\mathfrak{T}(\mathfrak{M})$ we denote the tangential fibering (tangent bundle) of the manifold $\mathfrak{M}$. By definition, $\mathfrak{T}(\mathfrak{M})$ is the set of all tangent planes $\mathfrak{T}_x$ with the natural smooth structure that converts $\mathfrak{T}(\mathfrak{M})$ into a manifold of class C^∞.

A section a of class C^r of the tangent bundle is called a vector field of class C^r on the manifold $\mathfrak{M} = \mathfrak{M}^n$. In other words, for every point $x \in \mathfrak{M}$ there is given a vector $a(x) \in \mathfrak{T}_x$ that depends smoothly on x. With the vector field a is it possible to construct a system of differential equations

$$\frac{dx}{dt} = a(x) \qquad (*)$$

The motion along the trajectories of this system occurs in such a way that the velocity vector of the particle located at the point x is equal to $a(x)$ and does not depend on t (autonomy).

In the future we will take $r = \infty$. Hence the existence and uniqueness theorems hold for the system of differential equations (*). Consequently, for every moment of time t and for every initial point x_0 there has been determined a point x_t, called the translation of the initial point x_0 by time t along the solution of the system.

We will examine the transformation S_t, which consists of the translation of a given origin along the trajectory after time t:

$$S_t : x_0 \to x_t .$$

The transformations S_t form a group in the obvious way: $S_{t_1} \circ S_{t_2} = S_{t_1+t_2}$. The groups $\{S_t\}$ arising from systems of differential equations are often called classical dynamical systems.

The local properties of trajectories are studied in the usual theory of differential equations. Ergodic theory is interested first of all in problems related to the behaviour of the trajectory as a whole.

Now we will examine the problem of the existence of a smooth invariant measure connected with the given system of differential equations. The theorem of Bogoliubov-Krilov yields, in the compact case, the existence of at least one invariant measure, but says nothing about its smoothness. If we are interested in the existence of a smooth invariant measure, then such a measure can be given by a differential form ω of dimension n (we assume that our manifold $\mathfrak{M}$ is oriented). In an arbitrary system of coordinates the form ω is given by a non-negative density function $\rho(x) : \omega = \rho(x) \prod_{i=1}^{n} dx_i$. Suppose that the components of the vector field $a(x)$ have the form $(a_1(x_1, \ldots, x_n), \ldots, a_n(x_1, \ldots, x_n))$ in the variables $x_1, x_2, \ldots, x_n$. Then our system of equations takes the form

$$\frac{dx_i}{dt} = a_i(x_1, \ldots, x_n) \qquad (**)$$

We have the fundamental

Liouville Theorem. In order that the differential form ω give rise to an invariant measure for the system $\{S_t\}$ it is necessary and sufficient that

$$\sum_{i=1}^{n} \frac{\partial(\rho a_i)}{\partial x_i} = 0 \quad \text{or} \quad \dim(\rho\vec{a}) = 0 .$$

A complete proof of this theorem can be found in the books "Qualitative theory of differential equations" by Nemitsky, Stepanov or "Analytical dynamics" by Whittaker or "Lectures on differential geometry" by Sternberg. The theorem of Liouville also yields the following Corollary: Let there be given a vector field $a(x)$ for which an invariant measure is given by a differential form ω. We examine the field $\beta(x)$ obtained from $a(x)$ "by a change in velocity":

$$\beta(x) = f(x)a(x), \quad f(x) > 0 .$$

Then β has a smooth invariant measure given by the differential form $\omega_1 = \frac{1}{f}\omega$.

The assertion in the corollary is checked by simple substitution of a in the Liouville theorem. As a matter of fact, it holds in the case of a time change in the abstract dynamical system. In the case $n > \infty$ for dynamical systems arising from variational mechanical principles, it is possible to find a smooth invariant measure.

Example. Hamiltonian systems.

Let $n = 2m$ and let the manifold $\mathfrak{M}^{2m}$ be a manifold of class C^∞. Assume that on $\mathfrak{M}^n$ there are introduced the coordinates which are partioned into two classes:

$q^1, q^2, \ldots, q^n$ (coordinates)

$p^1, p^2, \ldots, p^n$ (momenta)

Let there be given a function $\mathcal{H}(q,p)$ on $\mathcal{M}^{2m}$. The Hamiltonian system produced by the function $\mathcal{H}$ is the system of differential equations

$$\frac{dq^i}{dt} = \frac{\partial \mathcal{H}}{\partial p^i}, \quad \frac{dp^i}{dt} = -\frac{\partial \mathcal{H}}{\partial q^i}.$$

of the corresponding dynamical system, i.e. the group $\{S_t\}$. Many equations of classical mechanics can be written in that form.

To a hamiltonian system there corresponds a natural invariant measure, arising from the differential form

$$\omega = \prod_{i=1}^{n} dq^i dp^i \qquad (\text{i.e. } \rho(x) = 1).$$

This assertion follows immediately from the Liouville theorem and from the hamiltonian equations.

The measure that we have obtained bears the name of Liouville measure. The following difficulty is connected to it: The corresponding phase space is not compact and hence the Liouville measure is not finite on our space. There is a standard scheme to avoid this difficulty. We will examine a hamiltonian system with a hamiltonian function $\mathcal{H}$ that does not depend on time. Then $\mathcal{H}$ will be the first integral of this system, i.e. the total derivative of the function $\mathcal{H}$ with respect to time is equal to zero:

$$\frac{d\mathcal{H}}{dt} = \Sigma \frac{\partial \mathcal{H}}{\partial q^i} \frac{dq^i}{dt} + \Sigma \frac{\partial \mathcal{H}}{\partial p^i} \frac{dp^i}{dt} = 0 .$$

This means that the surface of constant energy $\mathcal{H}$ = const is invariant, i.e. every trajectory located on this surface at the initial moment $t = 0$ remains on it for all moments of time. The surface $\mathcal{H}$ = constant in the phase space, for typical values of the constant, often turns out to be already a smooth compact manifold on which a finite measure is induced. In statistical mechanics this induced measure is called a microcanonical distribution.

Finite dimensional linear systems of differential equations from the point of view of ergodic theory.

The hamiltonian function $\mathcal{H}$ is the energy of the system. In many cases the energy of the system presents itself as the sum of the kinetic and the potential energy.

The kinetic energy $\mathcal{H}_{kin}$ has, as a rule, the form

$$\mathcal{H}_{kin} = \Sigma\, a_{ij}(q)p^i p^j$$

where the $a_{ij}(q)$ are the coefficients of a positive definite quadratic form that depends only on the coordinates, and the p^i are the momenta. We take the potential energy $\mathcal{H}_{pot}$ in the form of a function $V(q)$ which depends only on the coordinates. The equilibrium positions of the system are those points for which $V(q)$ has an extremum. If $\bar{q}$ is an equilibrium position and $H_0 = V(\bar{q})$ then a point for which $q^i(0) = \bar{q}^i$, $p^i = 0$ remains immobile. If $\bar{q}$ is a point at a nondegenerate minimum then the corresponding equilibrium position will be stable.

For a stable equilibrium position $\bar{q}$ with $H = H_0 + \delta H$ all motions occur within a sufficiently small neighborhood of the point $\bar{q}$ for sufficiently small δH.

Hence the function $\mathcal{H}$ near the point $\bar{q}$ can be expanded in a series accurate up to second order terms in the variables p and $q - \bar{q}^i$

$$\mathcal{H} = \Sigma\, a_{ij}(\bar{q})p^i p^j + \Sigma\, V_{ij}(q-\bar{q})^i(q-\bar{q})^j \qquad (1)$$

The motions described by such a function $\mathcal{H}$ are called a linear approximation to the system around an equilibrium position.

If the function $\mathcal{H}$ has the form (1) then the solutions to the linear system are determined by the numbers a_{ij} and V_{ij}.

The quadratic form $\| a_{ij} \|$ is positive definite. By a well-known theorem in linear algebra there exists a change of coordinates

$$\tilde{q}^i = \Sigma\, c^i_j q^j, \quad \tilde{p}^i = \Sigma\, c^i_j p^j, \qquad i = 1, 2, \ldots, n$$

such that both forms are brought simultaneously to canonical form:

$$\mathcal{H} = \sum_{i=1}^{n} \lambda_i (\tilde{p}^i)^2 + \sum_{i=1}^{n} \omega_i (\tilde{q}^i)^2 .$$

Such a change of coordinates preserves the hamiltonian structure of the equations, i.e. the equations of motion have a hamiltonian form as before:

$$\frac{d\tilde{q}^i}{dt} = 2\lambda_i \tilde{p}^i , \quad \frac{d\tilde{p}^i}{dt} = -2\omega_i \tilde{q}^i .$$

It follows from this that for all $i = 1, 2, \ldots, n$ the equation below takes place:

$$\frac{d^2\tilde{q}^i}{dt^2} = -4\lambda_i\omega_i\tilde{q}^i \qquad \text{or} \qquad \frac{d^2\tilde{q}^i}{dt^2} + a_i\tilde{q}^i = 0, \; a_i = 4\lambda_i\omega_i .$$

For all i for which $\omega_i > 0$ we have $a_i > 0$.

The quantity $I_i = \omega_i(\tilde{q}^i)^2 + \lambda_i(\tilde{p}^i)^2$ serves as the first integral of the system. In that way, the linear systems have n independent first integrals $I_1, I_2, \ldots, I_n$. If $\omega_i > 0$ for all i then the intersection of the manifolds $I_1 = I_1^{(0)} = \text{const}, \ldots, I_n = I_n^{(0)} = \text{const}$ is compact and, as a rule, is an n-dimensional torus. This means that the linear hamiltonian systems are not ergodic and for $\omega_i > 0$ the phase space decomposes into a family of invariant n-dimensional tori. If the angular variables φ_i are introduced by the formula

$$\varphi_i = \arctan\left(2\lambda_i\sqrt{\frac{\lambda_i}{\omega_i}}\,\frac{\tilde{p}^i}{\tilde{q}^i}\right),$$

then the angular velocity $\frac{d\varphi_i}{dt} = a_i = \text{const}$, and we obtain a conditionally periodic motion on the n-dimensional torus of the form examined above. This system will be ergodic under rational independence of the numbers a_i.

<u>Ergodic theory of solutions of linear wave equations.</u>

Our previous analysis admits generalization to the infinite dimensional case, i.e. the case of a dynamical system given by partial differential equations. We will examine small oscillations of a string, membrane, or a bounded region in space. The equations of small oscillations can be obtained from a variational principle if in place of the hamiltonian function the following functional is taken:

$$\mathcal{H} = \frac{1}{2} \int \rho(x)\mu_t^2 \, dx + \frac{1}{2} \int k(x)(\nabla\mu)^2 \, dx .$$

In the future we will limit ourselves for simplicity to the one dimensional case and to problems with fixed boundary. The equations of motion of the string take the form

$$\rho(x) \frac{\partial^2 \mu}{\partial t^2} = \frac{\partial}{\partial x} (k(x) \frac{\partial \mu}{\partial x}) .$$

We will show how a dynamical system in functional space can be compared to this equation. In the role of phase space we will examine the space of pairs (μ, μ_t) where the functions μ, μ_t will later be subject to certain conditions. It follows from the Fourier method that if the $v_k(x)$ are eigenfunctions for the Sturm-Liouville problem for the equations

$$\frac{\partial}{\partial x}(k(x) \frac{\partial v_k}{\partial x}) = - \lambda_k \rho(x) v_k, \quad \text{where} \quad 0 \le \lambda_0 \le \lambda_1 \le \dots$$

and the functions μ, μ_t allow an expansion into a L^2 converging Fourier series with respect to the functions v_k convergent in L^2:

$$\mu = \Sigma\, a_k v_k ,$$

$$\mu_t = \Sigma\, b_k v_k ,$$

then the wave equation reduces to the following infinite system of ordinary differential equations:

$$\frac{da_k}{dt} = b_k, \quad \frac{db_k}{dt} = -\frac{1}{\lambda_k} a_k, \qquad k = 0, 1, \ldots$$

It follows from the form of this system that for the wave equation there is an infinite system of first integrals

$$a_k^2 + \lambda_k b_k^2 = I_k .$$

And so, we take for phase space $\mathcal{M}$ the space of pairs (μ, μ_t), where μ, μ_t can be expanded into Fourier series convergent in L^2. The fixing of the first integrals $I_0, I_1, \ldots$ distinguishes an infinite dimensional torus in the phase space. The motions of our system reduces to a conditionally periodic motion on that torus with frequencies $\omega_k = \pm\sqrt{\lambda_k}$. As in the finite dimensional case, it will be ergodic if any finite set of frequencies ω_k consists of rationally independent numbers. Apparently, for functions ρ, k of general form this will indeed be so, but rigorous proof of this fact is not known to us.

If in the functional space of pairs (μ, μ_t) the coordinates of "action" $I_0, I_1, \ldots$ and the "angle" $\varphi_0, \varphi_1, \ldots, \varphi_i = \arctan \frac{\sqrt{\lambda_i} b_i}{a_i}$ are introduced then every invariant measure for the dynamical system corresponding to the wave equation can be written in the form:

$$d\mu = d\lambda(I_0, I_1, \ldots) \cdot \prod_{k=0}^{\infty} d\varphi_k ,$$

where λ is an some measure in the space of variables $I_0, I_1, \ldots$.

Lecture 8

ERGODIC THEORY OF AN IDEAL GAS

Now we will examine the mathematical model of an ideal gas on the line (i.e. the one dimensional ideal gas).

Let there be given a circle of length L, on which there are distributed N points of mass 1, each of which moves with its constant velocity v_i. If q_i is the angular coordinate of the point i, then its equations of motion have the form

$$\frac{dq_i}{dt} = v_i ,$$

$$\frac{dv_i}{dt} = 0 .$$

The system of equations written above is a hamiltonian system with hamiltonian function $\mathcal{H} = \frac{1}{2}\sum_{i=1}^{N} v_i^2$. We denote

$$v_i = p_i ,$$

and so $\mathcal{H} = \frac{1}{2}\Sigma p_i^2$. Then, according to Liouville's theorem, there exists an invariant measure m which has the form

$$dm = \prod dq_i \, dp_i$$

The surface of constant energy $\mathcal{H} = \mathcal{H}(N, L) = \frac{1}{2}\sum_{i=1}^{N} p_i^2$ is the direct product of the N-dimensional torus (coordinate or configuration space) and the sphere of radius $\sqrt{2\mathcal{H}}$. We denote $d\tilde{\mu} = dq \times d\lambda$, where λ is the uniform measure on the sphere, and $dq = \prod_{i=1}^{N} dq_i$. The measure $\tilde{\mu}$ is clearly invariant.

The system under examination with data N, L, H examined by us is not ergodic, since the velocity of each particle is conserved. If the values of these velocities are fixed, then we will obtain a conditionally periodic motion on the N-dimensional torus. Since rational relations between components are encountered with probability 0 with respect to the measure $\tilde{\mu}$, such a system will be ergodic with probability 1.

Now we will perform an important transformation which will allow us to gain freedom from the numeration of the particles and execute a special passage to the limit $N \to \infty$ (thermodynamic limit). We pass from the original phase space to a new space, in which a set of N points of the circle serves as point (i.e. an N point subset of the circle). Earlier, the configuration space was $\mathrm{Tor}^N = \underbrace{S^1 \times S^1 \times \ldots \times S^1}_{N \text{ times}}$. Now we factor the space Tor^N by the group S_N, the group of permutations of N symbols. The set $Q_N = \mathrm{Tor}^N/S_N$ obtained will be a configuration space appearing as a pseudo-manifold, i.e., a manifold everywhere except for a subset of lower dimension, which is not frightening from the point of view of measure theory.

For every point $q \in Q_N$ the set of velocities of the m particles can be viewed as a function on q. If $q_i \in q$ then $v(q_i)$ is the velocity of the particle located at the point q_i. If there are k particles located at the point q_i then $v(q_i)$ is the k-tuple of the velocities of the particles located at q_i ordered by magnitude.

Now we will dwell on a construction that makes it possible to pass to the consideration of an ideal gas on the line with infinitely many particles. We

will suppose that there is given a sequence of circles having a fixed common point 0 and arranged as shown in the drawing. Let $N \to \infty$, $L \to \infty$, $\mathcal{H} \to \infty$, but the convergence to infinity does not take place in an arbitrary way, but is such that $\frac{N}{L} \to \rho$, $\frac{\mathcal{H}}{N} \to h$, where ρ and h are fixed positive numbers.

Under $L \to \infty$ the circles become "more straight" and in the limit we obtain a system with infinitely many particles on the line.

We will construct an invariant measure on the limit space as the limit of the invariant measures of Liouville, and we will also show that the limit system is ergodic and that it has even the mixing property. This will require the use of a somewhat more serious apparatus of measure theory.

Let Δ be an arbitrary interval on the line. We will denote by $\eta(\Delta)$ the number of points (molecules) that have fallen into the interval Δ. In the prelimit case (the length of the circle is L, the number of particles is N)

$$P_{N,L}\{\eta(\Delta) = k\} = C_N^k \left(\frac{\Delta}{L}\right)^k \left(1 - \frac{\Delta}{L}\right)^{N-k} .$$

Under our passage to the limit $P_{N,L}\{\eta(\Delta) = k\} \to e^{-\rho|\Delta|} \frac{(\rho|\Delta|)^k}{k!}$, where $|\Delta|$ is the length of Δ (Poisson distribution). Furthermore, the random variables $\eta(\Delta_1)$ and $\eta(\Delta_2)$ are independent in the limit if $\Delta_1 \cap \Delta_2 = \phi$.

These arguments show that the Poisson measure serves as the limit measure in the configuat ion space.

Now we will study the distribution of the velocities of the particles (see, for example, "Probability and related problems in Physics" by Kac, pp. 14-16). If $N < \infty$, $L < \infty$, then the N-dimensional velocity vector is uniformly distributed on the N-dimensional sphere of radius $\sqrt{2\mathcal{H}(N, L)}$,

$\sum_{i=1}^{N} v_i^2 = 2\mathcal{H}(N, L)$. It is not difficult to show, passing to a spherical coordinate system, that the distribution of each coordinate v_i converges:

$$\text{Distribution } v_i \xrightarrow{\text{weakly}} N(0, \sigma^2)$$

where $N(0, \sigma^2)$ is the Gaussian distribution with mean 0 and standard deviation σ^2, and $\sigma^2 = 2h$ and for different i the velocities are asymptotically independent. And so, the velocity of each particle has a Gaussian distribution with mean 0 and variance $\sigma^2 = 2h$, where the velocities of different particles are independent of each other (in this situation it is said that the probability distribution of the velocities of the gas is a Maxwell distribution). In the configuration space the particles are distributed according to the Poisson law with parameter ρ. It is easy to verify that the measure constructed is invariant with respect to the dynamics under which each particle moves with its velocity. We will show that the dynamics of an ideal gas are such that the corresponding dynamical system possesses the mixing property. Let x be a point of the phase space. Then $x = (q, v_q)$, where q is a countable subset of points of the line, and v_q is a function on that subset that assigns to each of its points (i.e. particles) its velocity.

Let Δ be a measurable subset of the line; for example an interval. Then it is possible to examine those subsets of the phase space $\mathfrak{M}$ which are defined by the position and the velocity of the particles that are located in that interval. An example of such a subset can be the set $A_k(\Delta)$, the set of x which contain precisely k particles in the interval Δ.

For every measurable set Δ the subsets of the phase space determined by it form a σ-subalgebra of the σ-algebra of all measurable subsets, which we denote by $\gamma(\Delta)$.

We note one important property of the subalgebra $\gamma(\Delta)$: If $\Delta_1 \cap \Delta_2 = \phi$ then the σ-algebras $\gamma(\Delta_1)$ and $\gamma(\Delta_2)$ are independent, i.e. for every set $A \in \gamma(\Delta_1)$ and for every set $B \in \gamma(\Delta_2)$

$$\mathcal{P}(A \cap B) = \mathcal{P}(A)\mathcal{P}(B)$$

We turn now to the proof of mixing. Mixing is equivalent to the following proposition: For every function defined on the space

$$\int f(S_t x) f(x)\, d\mu \xrightarrow[t\to\infty]{} \left(\int f\, d\mu\right)^2 .$$

It suffices to prove the mixing property for an everywhere dense set of functions in $\mathcal{L}^2(\mathcal{M}, \mu)$. For the role of such a set of functions we choose those functions that depend on the part of x in an interval Δ (measurable with respect to some σ-algebra $\gamma(\Delta)$).

Example 1. $f_\Delta(x)$ is the number of particles of the configuration x that have fallen into the interval Δ.

Example 2. $f_\Delta(x) = \sum_{q_i \in \Delta} v_q(q_i)$ is the total momentum of the particles located in the interval Δ.

Among all functions f we will examine those for which $\int f\, d\mu = 0$. In order to establish mixing it is necessary to show that

$$\int f(S_t x) f(x)\, d\mu(x) \xrightarrow[t\to\infty]{} 0 .$$

We will clarify the idea of the proof on the example of the function $f = f_\Delta$, where f_Δ is the number of molecules located in Δ. Then $f_\Delta(S_t x) = f(S_t x)$. We take a point x and a particle q_i that enters in it (moving with a velocity σ_i). After time t this particle will be located at the point $q_i + tv_i$ since there is no collision with other molecules (ideal gas!). We are interested in the probability of the event that the point $q_i + tv_i$ lies in Δ. If the point q_i was in Δ then under the condition $|v_i| \geq \frac{|\Delta|}{t}$ the particle will leave Δ, and as $t \to \infty$ the probability of that occurrence tends to one. And so, the probability of those x which contain particles that do not leave Δ after time t tends to zero as $t \to \infty$. But then $f(S_t x)$ becomes asymptotically measurable with respect to $\gamma(R^1 - \Delta)$. The independence of $\gamma(\Delta)$ and $\gamma(R^1 - \Delta)$ leads to the required relation.

Here is a more accurate proof. Let $f_1, f_2 \in \mathcal{L}^2(M,\mu)$ be bounded measurable with respect to the σ-algebra $\gamma(\Delta)$, Δ is some finite interval. For any $t > 0$ let us define a new function $f_1^{(t)}(x)$ by the following rule. Having the particles $(q_i, v_i) \in x$, $q_i \notin \Delta$ let us consider all particles for which $q_i + tv_i = q_i' \in \Delta$. The set of these particles $(q_i + tv_i, v_i)$ can be considered as part of the point of M inside Δ. We put $f_1^{(t)}(x)$ to be equal to $f_1(\cdot)$ where inside the brackets there is a set of particles $(q_i + tv_i, v_i)$, $q_i + tv_i \in \Delta$. The function $f_1^{(t)}(x)$ is measurable with respect to the σ-algebra $\gamma(R^1 - \Delta)$. If $S_t x = y$ is such that all particles inside Δ leave Δ for $q < -t$ then $f(S_t x) = f_1^{(t)}(x)$. We have

$$\lim_{t\to\infty} \int f_1(S_t x) f_2(x)\, d\mu = \lim_{t\to\infty} \int [f_1(S_t x) - f_1^{(t)}(x)] f_2(x)\, d\mu(x) + \lim_{t\to\infty} \int f_1^{(t)}(x) f_2(x)\, d\mu(x)$$

The first limit is equal to zero, because the difference $f_1^{(t)} - f_1(S_t x)$ tends to zero in probability and, therefore in $\mathcal{L}^2$ because f_1 is bounded. The second term is equal to zero because

$$\int f_1^{(t)}(x) f_2(x)\, d\mu(x) = \int f_1^{(t)}(x)\, d\mu(x) \int f_2(x)\, d\mu(x) = 0$$

in view of independence of σ-algebras $\gamma(\Delta)$, $\gamma(R^1 - \Delta)$. Our assertion is proved.

The information can be found in K. L. Volkovysky and Ya. G. Sinai "Ergodic Properties of an Ideal Gas with Infinitely Many Degrees of Freedon", "Functional Analysis and its Applications, vol. 5, No. 4, 1971, 19-21. See also S. Goldstein, Dynamical Systems with Infinitely Many Degrees of Freedom, Princeton dissertation.

Lecture 9

GEODESIC FLOWS ON RIEMANNIAN MANIFOLDS

Let Q be a compact, closed Riemannian manifold. This means that in each tangent space it is possible to introduce a scalar product $\langle e_1, e_2 \rangle$ that depends smoothly on the point q in the natural sense. In addition, for every smooth curve $\gamma \in Q$ it is possible to determine its length, it is possible to introduce on the manifold an element of area $d\sigma$ (Riemannian volume), admitting the integration of "good" functions: $\int f(q)\, d\sigma(q)$, and to introduce also parallel displacement of vectors and geodesic lines.

Let $\mathcal{M}$ be the unit tangent bundle on Q, i.e.

$$\mathcal{M} = \{x : (q, v) = x,\ q \in Q,\ v = v_q \in \mathcal{T}_q,\ \|v_q\| = 1\}$$

Then x is called the linear element and q is called the carrier of the linear element.

The set of linear elements corresponding to one and the same carrier forms a unit sphere S_q^{m-1}, where $\dim Q = m$. It is clear that $\dim \mathcal{M} = 2m - 1$.

Having the Riemannian metric on Q, it is possible to introduce a Riemannian metric on $\mathcal{M}$. Namely, let $x_1 = (q_1, v_1)$, $x_2 = (q_2, v_2)$ be two linear elements that are close together. For them we define $ds^2(x_1, x_2) = dq^2 + d\varphi^2$, where dq^2 is the square of the length of the geodesic connecting q_1 and q_2, and $d\varphi^2$ is the square of the distance between v_1 and the vector obtained from v_2 by parallel translation of it along that geodesic to the point v_1. We denote by μ the measure on $\mathcal{M}$ obtained from the metric ds.

We will examine the one-parameter group of transformations $\{S_t\}$ on the space $\mathcal{M}$ for which the individual transformation S_t consists of the translation of the linear element x along the geodesic determined by it through a distance t. Such a one-parameter group is called a geodesic flow.

Many important problems of classical mechanics lead to geodesic flows. A geodesic flow by itself admits a mechanical interpretation. Namely, we suppose that a material point of mass m, having energy h, moves on the surface of Q without friction. Then, under the appropriate choice of units, $\mathcal{M}$ will be the phase space of this dynamical system, and its motion on this space will be a geodesic flow. In a more general way, if the smooth manifold Q serves as coordinate space for some system, and the hamiltonian function is the the sum of the kinetic and potential energies, where the kinetic energy is a quadratic form of the momenta with coefficients depending on $q \in Q$, then it is possible to introduce a Riemannian metric on Q such that the motion of our dynamical system is obtained from the geodesic flow with smooth time change.

It is clear that, in the case of a smooth manifold, there corresponds a smooth vector field on $\mathcal{M}$ to the group $\{S_t\}$ of translations along geodesics. The applicability of ergodic theory to the study of geodesic flows is derived from the following lemma.

Lemma 1 (without proof). The geodesic flow preserves the measure μ. (See Russian Mathematical Surveys (Yenexu Mareuarurecrux Mayn) 1967, no. 5, the article of Anosov and Sinai, or "Lectures on classical mechanics" by V. I. Arnold).

Examples:

1. Let Q be the two dimensional sphere, then the geodesic lines are periodic on Q, the geodesic flow is not ergodic, and every geodesic is a separate ergodic component.

2. Let Q be a surface of revolution. This means that Q is obtained through the rotation of some curve in the plane (x, z) in space around the z axis.

In that case there is a one-parameter group of transformations derived from the rotation around the z axis acting on $\mathfrak{M}$. The orbits of this group are the closed curves consisting of linear elements, the carriers of which lie on a meridian, where the angle with the meridian is fixed. This group commutes with the geodesic flow, hence by Noether's theorem there is a corresponding integral of motion. This integral is called the Clairaut integral. From this point of view of mechanics the Clairaut integral is the projection of the angular momentum on the z axis.

Let $I(x)$, $x \in \mathfrak{M}$ be the Clairaut integral. For every fixed c the level sets $\mathfrak{M}_c = \{x : I(x) = c\}$ are invariant sets of the geodesic flow. For typical c this set is a two dimensional torus from the topological point of view.

As we will see later, for typical c the motion on such a torus can be reduced through the appropriate change of coordinates to the ergodic conditionally periodic motion on the torus. In this way, in the case of a surface of revolution the geodesic flow is not ergodic, and its typical ergodic components reduce to conditionally periodic motions on the two dimensional torus.

3. Flows on convex surfaces.

What was said above about surfaces of revolution relates, in particular, to the ellipsoid of revolution. It turns out that the situation for arbitrary tri-axial ellipsoids is the same as that for surfaces of revolution. (Jacobi "Lectures on dynamics").

For arbitrary convex surfaces Poincaré proposed the conjecture that there exist three closed geodesics on such surfaces. Poincaré himself proved the existence of one such geodesic on surfaces close to the ellipsoid. The complete solution to the problem was obtained by Lusternik and Shnirelman in the thirties. In good cases one or two of these geodesic is stable (in linear approximation). Under sufficiently general conditions the theory of Kolmogorov-Arnold-Moser is applicable, from which it follows that in this case the geodesic flow is not ergodic.

4. The geodesic flow on surfaces of constant negative curvature. Such surfaces are, by their geometric properties, completely unlike convex surfaces. This is reflected in the properties of the geodesic flows on these surfaces. We will prove the following theorem for that case.

Theorem (Hedlund-Hopf). The geodesic flow on a surface of constant negative curvature is ergodic.

Remark. Likewise it can be proved that the geodesic flow on such surfaces possesses good statistical properties: Mixing, central limit theorem for good functions, and exponential decay of correlation. The proofs of these properties are not simple, and we will not present them here.

Before passing to the proof of the theorem, we will dwell on the structure of surfaces of constant negative curvature. We will examine the Poincaré model of the Lobachevskian plane. As is known, in this model the Lobachevskian plane is the upper half-plane of the complex plane $\operatorname{Im} z > 0$ with the metric $ds^2 = \frac{1}{y^2}(dx^2 + dy^2)$.

The motions of the Lobachevskian plane are the fractional linear transformations that take the upper half-plane into itself.

The geodesic lines in the Poincaré model are either circles orthogonal to the real axis, which is called the absolute, or they are vertical halflines. We will also examine oriented geodesics. Two directed geodesics

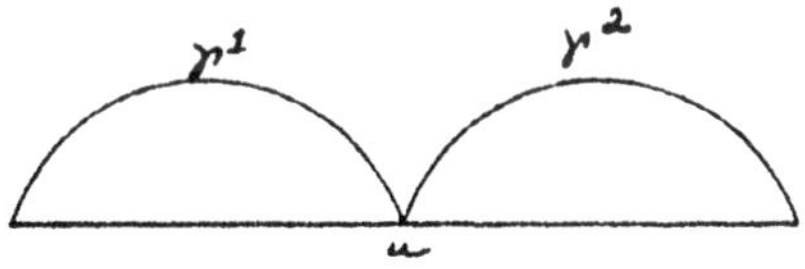

are called positively asymptotic if they end in the same point of the absolute. Let that point be u. We will determine the speed of contraction of such asymptotic geodesics as they tend to u. We perform a fractional linear transformation φ taking the upper half-plane into itself and the point u to infinity. Then $\varphi(\gamma^1)$ and $\varphi(\gamma^2)$ will be taken to two parallel lines intersecting the absolute at the points μ_1 and μ_2. The distance between the points of $\varphi(\gamma_1)$ and $\varphi(\gamma_2)$ located on the level $y = \nu$ $(\nu > 1)$, measured along the interval $\{y = \nu,\ \mu_1 \le x \le \mu_2\}$ equals

$$s = \int_{\mu_1}^{\mu_2} \frac{dx}{y} = \frac{\mu_2 - \mu_1}{y} \text{ ; and in}$$

turn $r = \int_1^{\nu} \frac{d\mu}{\mu} = \ln \nu$

i.e. $s = \frac{\mu_2 - \mu_1}{e^{\nu}}$

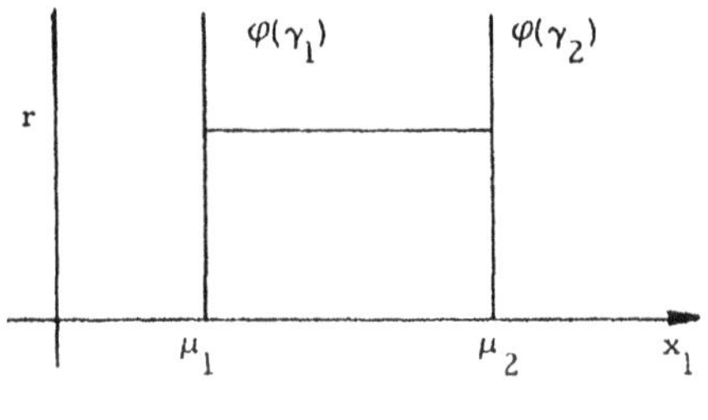

It follows from this that asymptotic geodesics approach each other with exponential speed. This property of asymptotic geodesics is the basis for the proof of ergodicity.

Fixing a point u of the absolute, we will examine the family of mutually asymptotic directed geodesics ending at the point u. It is clear that through every point of the plane there passes one and only one geodesic of that family. (If the geodesic lines are regarded as solutions of corresponding variational problems, then the family of directed asymptotic geodesics forms a field of extremals, according to the terminology used in the variational calculus).

We will examine the trajectories orthogonal to the family of asymptotic geodesics. In the Poincaré model the noneuclidean angles "coincide" with the euclidean angles, from which it follows that the orthogonal trajectories form circles that are tangent to the absolute at a point.

These circles are called horocycles.

In this way, to the family of directed asymptotic geodesics corresponds a family of orthogonal trajectories. A calculation similar to the above shows an important relation that holds between these two families. We take the interval $\tilde{\gamma}_0$ of the horocycle from a to b. Through each of its points c we construct a geodesic interval of length t in the direction of the point u. Then the union of the ends of these intervals is again an interval $\tilde{\gamma}_t$ of the horocycle, and their lengths are related by: $\ell(\tilde{\gamma}_t) = e^{-t}\ell(\tilde{\gamma}_0)$.

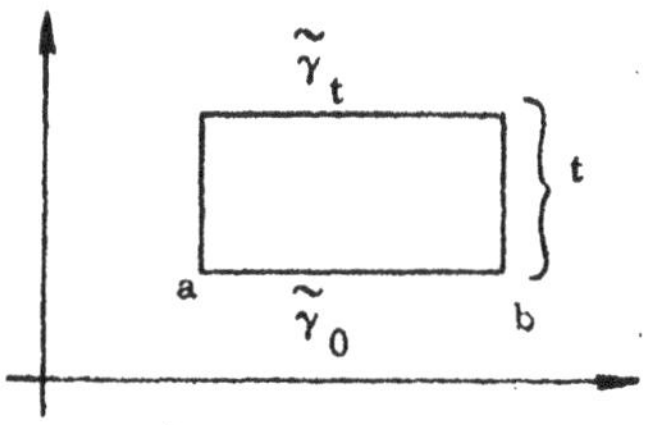

We "lift" these relations to the space of linear elements of the Lobachevskian plane. If Q_0 is the Lobachevskian plane then we denote by $\mathfrak{M}_0$ the space of linear elements of Q_0.

Every directed geodesic in Q_0 generates a curve in $\mathfrak{M}_0$: We have to examine the linear elements tangent to that geodesic and having positive direction on the given geodesic.

In an analogous way every horocycle in the Lobachevskian plane can be lifted to a curve in $\mathfrak{M}_0$. Namely, if x_0 is a linear element and q_0 is its carrier, then, through q_0, pass two circles $\tilde{\Gamma}^{(c)}$ and $\tilde{\Gamma}^{(e)}$ tangent to the absolute and orthogonal to x_0. We will examine the curves $\Gamma^{(c)}$ and $\Gamma^{(e)}$ in the space $\mathfrak{M}_0$ obtained when the curves $\tilde{\Gamma}^{(c)}$ and $\tilde{\Gamma}^{(e)}$ are equipped with unit normal vectors directed in the same way as x_0.

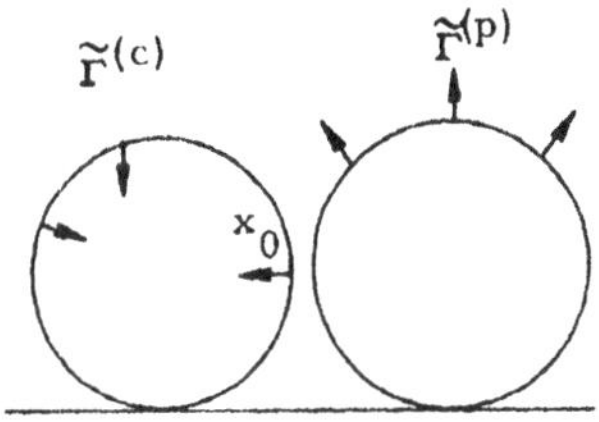

We will call the curves obtained in the space $\mathfrak{M}_0$ horocycles, as before. We note the following properties of horocycles:

1. $S_t\Gamma^{(c)}(x) = \Gamma^{(c)}(S_t x)$; $S_t\Gamma^{(e)}(x) = \Gamma^{(e)}(S_t x)$;

2. $\ell(S_t\gamma^{(c)}(x)) = e^{-t}\ell(\gamma^{(c)}(x))$; $\ell(S_{+t}\gamma^{(e)}(x)) = e^{t}\ell(\gamma^{(e)}(x))$;

$\gamma^{(c)}(x) \subset \Gamma^{(c)}(x)$ $\qquad$ $\gamma^{(e)}(x) \subset \Gamma^{(e)}(x)$.

(The proof is left to the reader).

In view of property 2, it is natural to call the horocycles $\Gamma^{(c)}$ and $\Gamma^{(e)}$ contracting and expanding, respectively, which determined the choice of indices "c" and "e" for them.

We denote by $R_\tau^{(c)}$ the transformation of the space $\mathfrak{M}_0$ consisting of the translation of every element $x_0 \in \mathfrak{M}_0$ along the contracting horocycle determined by it through a distance τ. It is easy to see that the transformation $R_\tau^{(c)}$ gives rise to a one-parameter group of smooth diffeomorphisms of the space $\mathfrak{M}_0$ and hence that it is generated by some smooth vector field which we denote $a^{(c)}$. The vector field $a^{(e)}$ for the horocycles $\Gamma^{(e)}$ is constructed analogously. We also denote by $a^{(0)}$ the smooth vector field corresponding to the geodesic flow. It is easy to see that at each point $x_0 \in \mathfrak{M}_0$ the vector fields $a^{(e)}, a^{(c)}, a^{(0)}$ are linearly independent among themselves.

Each compact surface of constant negative curvature is obtained from the Lobachevskian plane Q_0 with the aid of some discrete group of motions Γ. Namely, the surface Q is obtained as a quotient space Q_0/Γ. (See Springer "Introduction to the theory of Riemann surfaces").

Let $\mathfrak{M}$ be the space of linear elements of the surface Q, and let the natural transformation $Q_0 \to Q$ generate a transformation $\mathfrak{M}_0 \to \mathfrak{M}$. As before we will write $a^{(0)}, a^{(c)}, a^{(e)}$ for the vector fields on $\mathfrak{M}$ that are induced through the transformation by the corresponding fields on $\mathfrak{M}_0$. In particular, it follows from this, that for every $x_0 \in \mathfrak{M}$ there exist expanding and contracting horocycles passing through it.

We pass directly to the proof of the Hedlund-Hopf theorem. Let f(x) be a continuous function. By the ergodic theorem of Birkhoff-Khinchin, the following limits exist almost everywhere:

$$\lim_{T\to\infty} \frac{1}{T}\int_0^T f(S_t x)\,dt = \hat{f}^+(x), \tag{*}$$

$$\lim_{T\to\infty} \frac{1}{T}\int_0^T f(S_{-t} x)\,dt = \hat{f}^-(x), \tag{**}$$

and $\hat{f}^+ = \hat{f}^-$ almost everywhere on $\mathfrak{M}$.

We wish to prove that $\hat{f}^-$ = const almost everywhere. It is clear that it suffices to prove this assertion in a small neighborhood of every point x_0. Fixing this neighborhood and denoting it by U, we take a point $x \in U$ for which $\hat{f}^+(x)$ exists. We remark that:

1. If $\hat{f}(x)$ exists and $y \in \Gamma^{(c)}(x)$ then

$$\hat{f}^+(x) = \hat{f}^+(y) .$$

Indeed, this follows from the fact that the points $S_t x$ and $S_t y$ approach each other with exponential speed, i.e., $\hat{f}^+(x)$ is constant on contracting horocycles.

2. Analogously, if $\hat{f}^-(x)$ exists and $y \in \Gamma^{(e)}(x)$ then $\hat{f}^-(x) = \hat{f}^-(y)$. i.e., $\hat{f}^-(x)$ is constant on expanding horocycles.

We take some interval $\gamma^{(c)}$ of a contracting horocycle for which $\hat{f}^+ = \hat{f}^-$ almost everywhere. Since $\hat{f}^+ = \hat{f}^-$ almost everywhere in U, such intervals form a set of full measure in U. We construct a two dimensional surface $\ell^{(c)} = \bigcup_{|t|<t_0} S_t\gamma^{(c)}$ for which $\hat{f}^+ = \hat{f}^-$ almost everywhere as before.

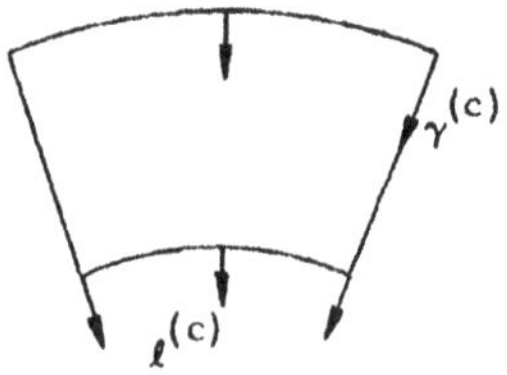

Since the vector field $a^{(e)}$ is smooth and $\ell^{(c)}$ is part of a smooth submanifold in $\mathcal{M}$, it is possible to choose a subset of full measure $A \subset \ell^{(c)}$ on $\ell^{(c)}$, where $A = \{z : z \in \ell^{(c)}, \hat{f}^{+}(z) = \hat{f}^{-}(z)\}$. Then

$$B = \bigcup_{z \in A} \gamma^{(e)}(z)$$

intersected with U yields a subset of full measure in U since the field $a^{(e)}$ is smooth and $\ell^{(c)}$ is part of a smooth surface. For every pair of points $x_1 \in B$, $x_2 \in B$ and their corresponding $z_1 \in A$, $z_2 \in A$ we have $\hat{f}^{-}(x_1) = \hat{f}^{-}(z_1) = \hat{f}^{+}(z_1) = \hat{f}^{+}(z_2) = \hat{f}^{-}(z_2) = \hat{f}^{-}(x_2)$. In this way, $\hat{f}^{-}$ is constant almost everywhere in U. Consequently, the geodesic flow is ergodic. The theorem is proved.

There is a vast literature devoted to ergodic properties of geodesic flows on manifolds of negative curvature and their generalization (so-called Anosov systems). The reader can find many deep results of topological and ergodic character in papers

1. S. Smale, Smooth, dynamical Systems., Bull. Amer. Math. Soc.
2. D. V. Anosov, Ja. G. Sinai, Some smooth ergodic dynamical systems. Uspehi Math. Nauk, 1967, v. 22, N5, 107-172.
3. D. V. Anosov, Geodesic Flows on Closed Riemannian Manifolds with Negative Curvature. Proc. of Steklov Inst. N. 90(1967).
4. R. Bowen, Equilibrium States and Ergodic Theory of Anosov Diffeomorphisms Lecture Notes in Math. No. 470, Springer Verlag 1975.

Lecture 10

BILLIARDS

Let Q be a bounded region in $\mathbb{R}^k$ with piece-wise smooth boundary $(K \geq 2)$.

Billiards in Q is the dynamical system generated by the uniform linear motion of a material point inside Q with a constant velocity and with reflection at the boundary such that the tangential component of the velocity remains constant and the normal component changes sign.

The phase space of a billiard consists of all possible pairs (q, v) where $q \in Q$ and v is the velocity vector, i.e. an element of the unit sphere in K-dimensional space. In that case q is called the carrier of the point x.

We will give the measure on phase space in the form

$$d\mu = dq \cdot d\omega ,$$

where $d\omega$ is the uniform distribution on the unit sphere of the K-dimensional space of points $x = (q, v)$ having carrier q. The measure μ is finite, and we will assume that it is normalized.

Those trajectories which lead to the singular points of the boundary have measure zero. From the point of view of ergodic theory, these trajectories can be neglected.

Assertion. The measure μ is an invariant measure for the billiards in Q.

Our assertion is evident inside the region. We will examine the reflection from the boundary. A reflection with respect to a plane preserves measure in an obvious way. The general case differs from the case of the plane in small higher order terms (since the boundary is smooth).

Let $\mathfrak{M}$ be the set of those linear elements (i.e. points $x = (q, v)$), whose carriers either lie in Q or belong to the boundary, where in the last case v points out of Q. Then it is possible to examine the transformation S_t taking the set $\mathfrak{M}$ into itself; $S_t : \mathfrak{M} \to \mathfrak{M}$ consisting of the change in position of the particle and of the direction of its velocity in accordance with its motion along the trajectory for time t.

$\{S_t\}$ will be a dynamical system on the space $\mathfrak{M}$.

Examples of billiards and questions related to them.

I. $K = 2$, Q is a rectangle.

Along a given trajectory of the billiard that does not pass through a vertex of the rectangle, the angle of reflection (i.e. the angle that the reflected trajectory makes with the interior normal) can assume only four possible values. Hence the billiards in a rectangle is not ergodic.

We pave the entire plane with rectangles with sides a and b and we

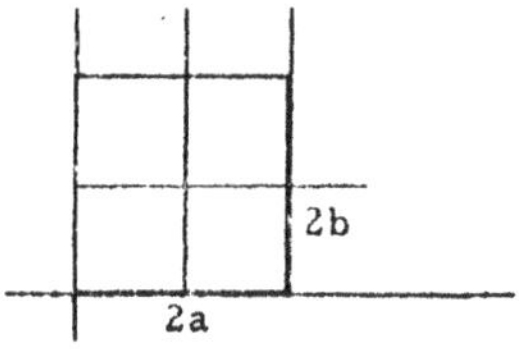

construct a transformation which we denote by π, of the plane to a rectangle with sides a and b. Let (x, y) be a point of the plane for which $x = Ka + x_0$, $y = \ell b + y_c$, where K, ℓ are integers and x_0, y_0 are such that $0 \leq x_0 < a$, $0 \leq y_0 < b$. We define the transformation $\pi(x, y)$ in the following way:

1) If K, ℓ are even then $\pi(x, y) = (x_0, y_0)$;

2) if K is even and ℓ is odd then $\pi(x, y) = (x_0, b - y_0)$;

3) if K is odd and ℓ is even then $\pi(x, y) = (a - x_0, b)$;

4) if K, ℓ are odd then $\pi(x, y) = (a - x_0, b - y_0)$.

It is easy to see that the transformation π is continuous and that the lifting of every billiard trajectory from the rectangle $Q = (a, b)$ to the plane is a straight line.

The identification shown above acts like the construction of the torus from the rectangle (2a, 2b). Hence the billiards in this rectangle can be examined as a linear motion on the two dimensional torus if the angle of inclination φ of the first interval of the trajectory is fixed.

For such a motion on the torus we have the condition of ergodicity: It is necessary that $\frac{a}{b} \tan \varphi$ be irrational.

II. The equilateral triangle.

The Billiards in an equilateral triangle is examined completely analogously to the billiards in a rectangle since the entire plane can also be paved with equilateral triangles.

III. The Billiards in an arbitrary triangle.

We will show that, if the angles of a triangle are rationally dependent billiards inside such a triangle is not ergodic. For that we will examine the reflection from all three sides of the triangle. We fix the initial direction from which we will measure the angle φ, $-\pi < \varphi \leq \pi$. Suppose, for definiteness, that it coincides with AB. If the particle moves with an angle φ_0 and reflects off AB then its angle after reflection equals $-\varphi_0 : \varphi_0 \to -\varphi_0$. If it moves with an angle φ_0 and reflects off the side AC then its angle after reflection equals

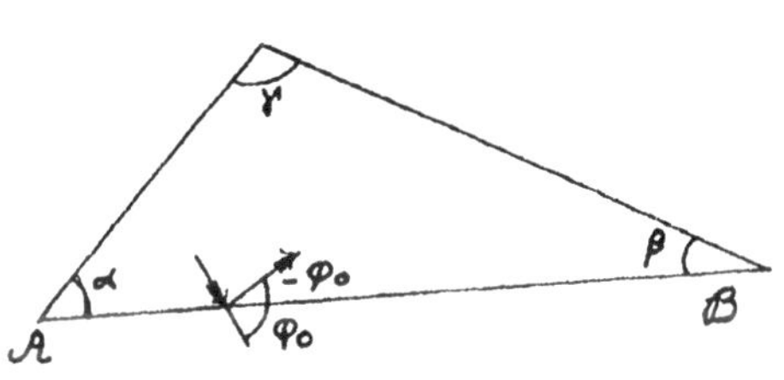

$\varphi_0 \to 2\alpha - \varphi_0$. If, finally, the particle moves with an angle φ_0 and reflects off the side BC then $\varphi_0 \to 2(\pi - \beta) - \varphi_0$. In this way, if $\alpha = 2\pi\frac{p}{q}$, $\beta = 2\pi\frac{r}{s}$, where p, q, r, s are integers, then for all possible reflections of the particle from the sides of the triangle the angle assumes only a finite set of values. It follows immediately from this that the motion is not ergodic.

Let $(\varphi_1, \varphi_2, \ldots, \varphi_k)$ be that finite set of angle values that we just talked about. Now we will show how a transformation of the permutation type arises naturally here.

$\varphi_1, \varphi_2, \ldots, \varphi_k$ are angles corresponding to linear elements, the carriers of which lie on the lower side of the triangle. If they are considered as intervals of corresponding length, then after the first reflection there appears a transformation of these intervals - a permutation. There corresponds a permutation to the billiards inside the triangle with rationally commensurable sides. At the present moment the conditions under which this permutation is transitive are not known.

Certain problems of mechanics reduce to the study of billiards. We will examine one of them.

Let there be 2 material points with masses m_1 and m_2 with velocities v_1 and v_2, respectively, moving on the interval $[0, 1]$. Suppose, in addition, that they rebound elastically from each other and from the endpoints of the interval. We denote by q_1 and q_2 the coordinates of the first and second points on the interval. Then $0 \leq q_1 \leq q_2 \leq 1$. The configuration space of our system is a triangle, and the trajectory of motion of the particles is a trajectory of the billiards type in this triangle. For the proof of the

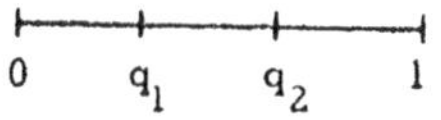

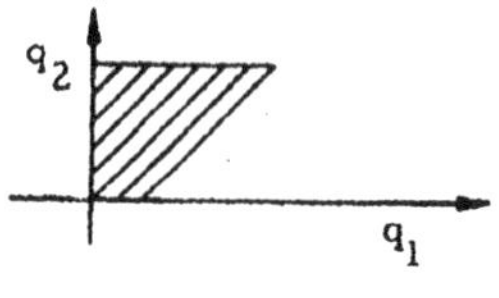

last assertion we introduce new coordinates: $q_1' = \sqrt{m_1}\, q_1$, $q_2' = \sqrt{m_2}\, q_2$. Now $0 \le q_1' \le \sqrt{m_1}$, $0 \le q_2' \le \sqrt{m_2}$, and the configuration space takes the form of a triangle: $0 \le q_1' \le \sqrt{\frac{m_1}{m_2}}\, q_2' \le \sqrt{m_1}$. If velocities corresponding to

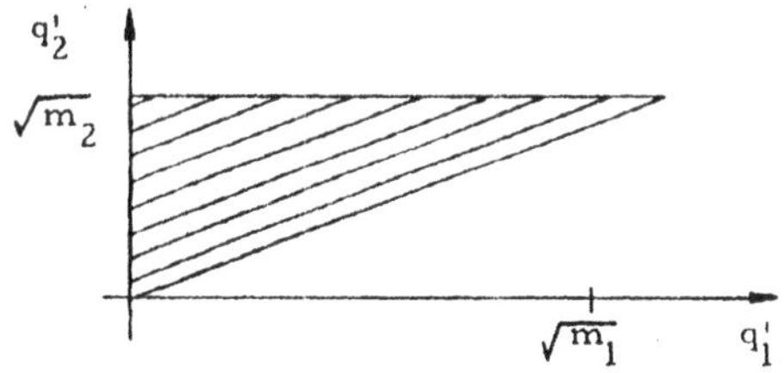

the coordinates q_1', q_2' are introduced, $v_1' = \sqrt{m_1}\, v_1$, $v_2' = \sqrt{m_2}\, v_2$, then the energy of the system becomes

$$H = \frac{(v_1')^2}{2} + \frac{(v_2')^2}{2} .$$

It follows from the law of conservation of energy that the length of the velocity vector (v_1', v_2'), equal to $\sqrt{2H}$, is preserved under reflections.

The component of the momentum of the particles along the hypotenuse is conserved in the collision: $\sqrt{m_1}\, v_1' + \sqrt{m_2}\, v_2' = \text{const}$. This shows that the scalar product of the velocity vector (v_1', v_2') with the hypotenuse vector $(\sqrt{m_1}, \sqrt{m_2})$ is constant. This means that the law of reflection from the hypotenuse is the law of reflection of a billiard. The reflection off the legs of the triangle is investigated even more simply, and we do not pursue it. If m_1 and m_2 are such that the angles α and β of the triangle can be presented in the form $\alpha = 2\pi \frac{p}{q}$, $\beta = 2\pi \frac{r}{s}$, where p, q, r, s are integers, then the motion is not ergodic. The question about the ergodicity for incommensurable α, β remains open.

The Billiards in convex regions.

Let the region Q be an ellipse with foci F_1, F_2.

Theorem. Every trajectory of the billiards on the ellipse is tangent to either a hyperbola or an ellipse confocal with the given ellipse.

Proof. We will prove this assertion for the case of the ellipse. Let A be a point of Q and let A_1A, AA_2 be two consecutive intervals of some trajectory of our billiard ball. We take the point B_1 that is symmetric to F_1 with respect to A_1A and we construct the interval B_1F_2. This interval intersects A_1A in the point C_1. It is clear that $\mu_1 = F_1C_1 + C_1F_2 = B_1C_1 + C_1F_2$ and that the inequality $\mu_1 < F_1D_1 + D_1F_2$ holds for every point D_1 of the interval A_1A distinct from C_1. It follows from this that the ellipse that is confocal with Q and for which the sum of the distances to F_1 and F_2 equals μ_1 is tangent to A_1A at the point C_1. We make an analogous construction with the interval AA_2. We find the corresponding point C_2 and $\mu_2 = F_1C_2 + C_2F_2$. We remark now that $\mu_1 = \mu_2$ in view of the congruence of the triangles B_1AF_2 and B_2AF_1. Consequently, C_1 and C_2 lie on the same ellipse. Our assertion is proved. A hyperbola arises in the case when the angle A_1AA_2 lies inside the angle F_1AF_2.

Let $\frac{x^2}{a^2-\mu} + \frac{y^2}{b^2-\mu} = 1$ be the equations of the ellipses and hyperbolas that are confocal with Q. Then for every set U, those billiard trajectories that are tangent to the confocal curves with $\mu \in U$ form an invariant set. Hence billiards in the ellipse is not ergodic.

Let Q be an arbitrary convex region, bounded by a curve Γ.

Definition. A caustic for billiards in Q is a curve γ such that the trajectory of the particle is tangent to the curve after each reflection.

Finding caustics is important for a series of asymptotic problems in the theory of partial differential equations. Recently the Leningrad mathematician V. F. Lazutkin proved that there exist many caustics in convex sets of general form. In addition, the measure of the set of linear elements that are tangent to caustics is positive, and the boundary of the billiards table serves in a natural way as a point of density of this set.

The converse problem is much simpler: Let there be given a convex curve γ. To find all curves Γ for which γ is a caustic. The student Minasian of Erevan University noticed that such curves Γ form a one-parameter family, and that they can be obtained in the following way: Let $s \geq \ell(\gamma)$, where $\ell(\gamma)$ is the length of γ. Then Γ consists of all points q such that if intervals tangent to γ are drawn from q and the sum of the lengths of these intervals and of the part of the curve γ that lies between their ends is taken, then it will be constant and equal to s.

A periodic trajectory of the billiards is a polygon in Q which may serve as a trajectory of the billiards in Q.

Theorem. Inside a smooth convex curve there always exist infinitely many periodic billiard trajectories.

The theorem and its proof, which we will present, belong to Birkhoff.

Proof. For an arbitrary $n \geq 3$ we will construct a periodic trajectory having n vertices. The following argument can be easily checked also for the case $n = 2$ as well. Let n be fixed. We will examine all possible convex n-agons π_n inscribed in the curve Γ. By $f(\pi_n)$ we denote the perimeter of the polygon π_n. It is possible to define the distance between any two n-agons inscribed in Γ, for example, as the distance between their sets of vertices. Then the set of all Π_n is a compact, closed, and complete metric space, and f is a continuous function on it. Hence, for some n-agon $\Pi_n^{(0)}$ it achieves its maximum value: $f(\Pi_n^{(0)}) = \max_{\pi_n} f(\Pi_n)$. We will prove that the boundary of $\Pi_n^{(0)}$ is a periodic trajectory for the billiard inside Γ.

Let A_1, A_2, and A_3 be three consecutive vertices of the polygon $\Pi_n^{(0)}$. We will examine the one parameter family of ellipses with foci at A_1 and A_3.

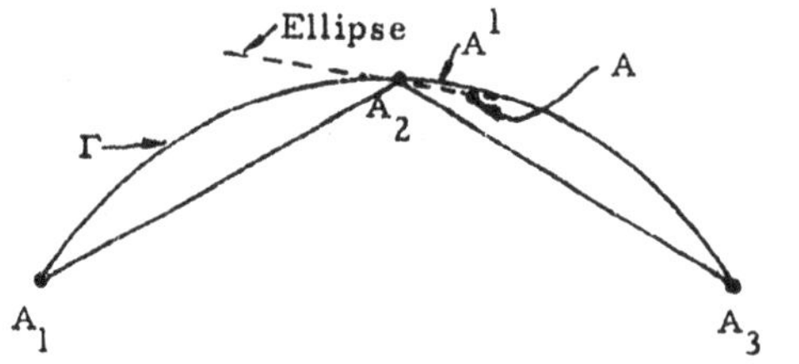

The ellipse that passes through A_2 and belongs to the family must be tangent to Γ at A_2. For suppose that it cuts Γ as shown in the figure. Then the sum $A_1A + AA_3$ is equal to the sum $A_1A_2 + A_2A_3$ and therefore there are points A^1 on Γ near A for which $A_1A^1 + A^1A_3 > A_1A_2 + A_2A_3$. This contradicts the defining maximality property of $\Gamma_n^{(0)}$. Since the ellipse is tangent to Γ at A_2, $A_1A_2A_3$ is a billiards trajectory. The theorem of Birkhoff is proved.

Recently A. B. Katok proved for the billiards in a triangle with comensurable angles that the dynamical system has no mixing on every ergodic component.

See

1. V. F. Lazutkin. The existence of caustics for billiards in convex domains. Izvestia Acad. of Sci. Ser. Math. 1973, V. 37, nl, 186-216.

2. M. M. Dvorin, V. F. Lazutkin. The existence of infinitely many elliptic and hyperbolic periodic trajectories for billiards in convex domains. Functional Analysis and Applications, 1973, V7, n 2, 20-27.

Lecture 11

DYNAMICAL SYSTEMS ON THE TWO DIMENSIONAL TORUS

Let T^2 be the two dimensional torus and let (x, y) be coordinates on T^2. The dynamical system that interests us is the one-parameter group of translations along the trajectories of the system of differential equations:

$$\frac{dx}{dt} = F_1(x, y)$$
$$\frac{dy}{dt} = F_2(x, y) \tag{1}$$

with invariant measure whose density equals $p(x, y)$. We will assume that the vector field of the system (1) does not have singular points, i.e. $F_1^2 + F_2^2 > 0$ and the functions $F_1, F_2, p \in C^\infty$. The problem of studying such systems appears in many problems, for example, in the study of geodesic flows on surfaces of revolution or Liouville surfaces, where the geodesic flow is integrated. For a fixed value of the first integral, in the three dimensional space of linear elements the two dimensional torus is distinguished, on which the geodesic flow induces a vector field of the form examined.

Theorem. There exists an infinitely differentiable change of coordinates:

$$u = u(x, y), \qquad v = v(x, y)$$

on the torus T^2 such that the trajectories of our system (1) in the new coordinate system are straight lines, and the system is written in the form

$$\frac{du}{dt} = F$$

$$\frac{dv}{dt} = \gamma F$$

where γ is a number and F is some positive function. In the proof we will make an additional assumption:

$$F_1(x, y) > 0, \quad (x, y) \in T^2$$

We will break the proof into separate parts.

1. There exists a time change such that for the new dynamical system the meridian $x = 0$ is transformed into the meridian $x = 1$ (i.e. into itself) after time $t = 1$.

Indeed, the system

$$\frac{dx}{dt} = 1$$
$$\frac{dy}{dt} = \frac{F_2}{F_1} \tag{2}$$

has the same trajectories as the system (1), and the time of transition of each point of the meridian $x = 0$ to the meridian $x = 1$ is clearly equal to one.

2. We will deal with the system (2).

After the time change the density of the invariant measure changes and becomes equal to (see lecture 7)

$$\rho_1(x, y) = \rho(x, y)F_1(x, y)$$

3. Let $\{S_t\}$ be a one-parameter group of translations along the trajectories of system (2). Applying S_t to the meridian $\gamma_0 = \{x = 0\}$, we obtain a smooth partition of the torus into meridians $S_t\gamma_0$, $0 \leq t \leq 1$,

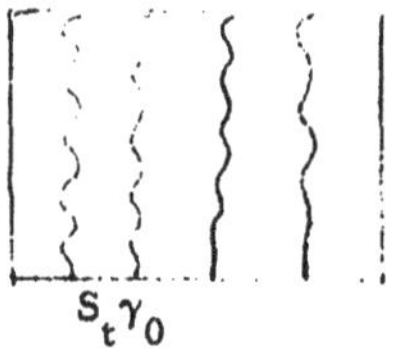

We will examine the circle $\gamma_0 = \{(x, y): x = 0\}$ and the transformation $T : \gamma_0 \to \gamma_0$, where $Ty_0 = y_1$ if $S_1(0, y_0) = (1, y_1)$. Then T is a diffeomorphism of the circle having an invariant measure with density $r(y_i)$, and it follows from the smoothness condition and the invariance that

$$r(y_0)\, dy_0 = r(Ty_0)T(dy_0) \qquad (*)$$

We will show that $r(y_0) = \rho_1(y_0, 0)$. For that it is necessary to prove that $\rho_1(y_0, 0)$ satisfies condition (*).

We will examine a small rectangle a around the point y_0 with sides parallel to the coordinate axes and equal to $\frac{1}{n}$ and dy_0, respectively.

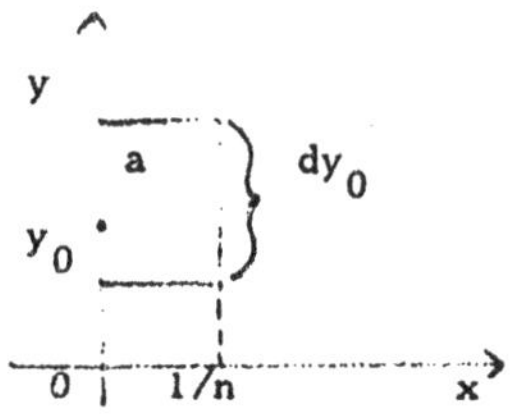

After a unit of time a will be transformed into a curvilinear parallelogram a', and it follows from the fact that S_t preserves measure that $\mu(a) = \mu(a')$. But for small dy_0 and $\frac{1}{n}$

$$\mu(a) = dy_0, \frac{1}{n}(\rho_1(y_0, 0)+0(1))$$

and hence, $\mu(a') = T(dy_0)\frac{1}{n}(\rho_1(Ty_0, 0)+0(1))$, consequently, $dy_0\rho_1(y_0, 0) = Tdy_0 \cdot \rho_1(Ty_0, 0)$, and $\rho_1(y_0, 0)$ satisfies condition (*).

4. For the diffeomorphism T of the circle, which preserves a smooth measure, there exists a coordinate system in which T is a rotation. We set $v(y_1, y_2) = \int_{y_1}^{y_2} r(u)\,du$ and we let $\alpha = \int_0^{T(0)} r(u)\,du$. Then

$$v(0, Ty_0) = v(0, T(0)) + v(T(0), Ty_0) = \alpha + v(T(0, y_0)) = \alpha + v(0, y_0) .$$

Having taken $v(0, y_0)$ as new coordinate for the point y_0, we obtain that T is a translation by α in that coordinate.

5. We will construct a coordinate system in which system (2) will be a system with constant coefficients.

We assume $u = x$. Let $v = \mathbf{v}$ for $x = 0$

$$v(x, y) = \mathbf{v}(0, y) + \alpha x \qquad \text{if} \qquad (x, y) = S_\alpha(0, y_0) .$$

In the new coordinate system (u, v) the trajectories are straight lines. Consequently, the trajectories of the original system are straight lines.

Theorem 1 is proved.

Theorem 1 shows that it is possible to restrict ourselves to the study of systems of the form

$$\frac{du}{dt} = F(u, v)$$
$$\frac{dv}{dt} = aF(u, v) \qquad (3)$$

Now we will examine the question when the system (3) can be reduced to a system with constant coefficients by a suitable change of coordinates. In order to obtain a system with constant coefficients it is sufficient that there exist a closed curve γ such that each of its points returns to that curve at one and the same time. We will construct an equation for the curve γ. Let v_0 be the coordinate on the meridian $\{x=0\}$ that was constructed in part 4, and let $\tau(v_0)$ be the time of motion of the point $(0, v_0)$ to the sought curve γ. It is clear that the function $\tau(v_0)$ can be looked at as a definition of the curve γ. We set

$$t(v_0) = \int_0^1 \frac{du}{F(u, v_0)}.$$

Then the equations for $\tau(v_0)$ have the appearance

$$t(v_0) - \tau(v_0) + \tau(v_0+a) = K \qquad (4)$$

It is seen immediately from equation (4) that $K = \int_0^1 t(v_0)dv_0$. We expand the functions $t(v_0)$ and $\tau(v_0)$ in their fourier series. We notice that

$$t(v_0) = \sum_{-\infty}^{\infty} t_k e^{2\pi i k v_0}, \qquad t_0 = K$$

$$\tau(v_0) = \sum_{-\infty}^{\infty} \tau_k e^{2\pi i k v_0}$$

Then equation (4) can be rewritten in the form

$$\sum_{k \neq 0} (t_k - \tau_k + \tau_k e^{2\pi i k \alpha}) e^{2\pi i k v_0} \equiv 0 \tag{4'}$$

or

$$\tau_k = \frac{t_k}{1 - e^{2\pi i k \alpha}}$$

It is necessary for the existence of the function τ that $\sum_{k=-\infty}^{\infty} |\tau_k|^2 < \infty$, i.e. for rational α equation (4) is in general nonsolvable. We will examine the question of when $\sum_{k=-\infty}^{\infty} |\tau_k|^2 < \infty$. We call the number α a normally approximated (by rationals) number if: There exist $C, \varepsilon > 0$ such that for every integer q

$$\min_p \left| \alpha - \frac{p}{q} \right| \geq \frac{C}{q^{2+\varepsilon}},$$

where the minimum is taken over all integers p.

<u>Lemma</u>. The normally approximated numbers α on the interval $[0, 1]$ form a set of full measure.

Indeed, fixing ε and C, we have $\mathrm{mes}(A_q) \leq \frac{2C}{q^{1+\varepsilon}}$, where

$$A_q = \{\alpha : \min_p |\alpha - \frac{p}{q}| < \frac{C}{q^{2+}}\} , \quad \Sigma_q \mathrm{mes}(A_q) < \infty .$$

The assertion of the lemma follows from this with the help of the Borel-Cantelli lemma.

We will evaluate τ_k for α under the condition that α is normally approximated.

By virtue of our evaluations we have: $\frac{1}{2k} \geq |\alpha - \frac{n}{k}| \geq \frac{C}{k^{2+\varepsilon}}$ or $\frac{1}{2} \geq |k\alpha - n| \geq \frac{C}{k^{1+\varepsilon}}$, hence $|1 - e^{2\pi i k\alpha}| \geq \frac{C_1}{k^{1+\varepsilon}}$; from which we conclude that $|\tau_k| < \frac{|t_k|}{C_1}, k^{1+\varepsilon}$.

<u>Theorem.</u> If $t(v_0) \in C^\infty$ and α is normally approximated then equation (4) is solvable and $\tau(v_0) \in C^\infty$.

<u>Proof:</u> Since $t(v_0) \in C^\infty$, we have $\frac{d^r t}{dv_0^r} = \Sigma_k (2\pi i k)^r t_k e^{2\pi i k v_0}$, and $|t_k| \leq \frac{\mathfrak{D}_r}{(2\pi)^r \cdot k^r}$, where $\mathfrak{D}_r = \max_{v_0 \in [0,1]} |\frac{d^r t(v_0)}{dv_0^r}|$. But then

$$|\tau_k| \leq |t_k| \cdot \frac{k^{1+\varepsilon}}{C_1} \leq \frac{\mathfrak{D}_r}{(2\pi)^r \cdot C_1 \cdot k^{r-1-\varepsilon}} ,$$

and $\tau(v_0) \in C^\infty$. The theorem is proved.

Having obtained $\tau(v_0)$, it is possible to transform it smoothly to the meridian $x = 0$, and also, applying the points 2, 3, 4 of theorem 1, to reduce the system to the form

$$\frac{du}{dt} = 1$$

$$\frac{dv}{dt} = \alpha$$

The above discussion is based on the work of A. N. Kolmogorov "On dynamical systems with integral invariance on the torus" DAN, vol. 93 763-766 (1953) and the book S. Sternberg's "Celestial Mechanics", Part I.

Lecture 12

DYNAMICAL SYSTEMS ARISING IN THE THEORY OF PROBABILITY

We will examine a random process with discrete time, i.e. a sequence of random real-valued variables

$$\ldots \xi_{-n}, \xi_{-n+1}, \ldots, \xi_{\infty} \xi_1, \ldots, \xi_n, \ldots = \{\xi_i\}_{-\infty}^{\infty},$$

where ξ_i is the result of the observation of a random object at the moment in time i, $-\infty < i < \infty$. In keeping with the tradition observed in the theory of probability, we will denote by Ω the space of all possible realizations of the random process. Let there be given a probability distribution for the random process, i.e. all probabilities of the form $P\{\xi_{i_1} \in A_1, \xi_{i_2} \in A_2, \ldots, \xi_{i_r} \in A_r\}$, where A_j is an arbitrary measurable set, are defined.

We will examine the transformation T, given on the space Ω, which consists of the following: If $\xi - \{\xi_i\}_{-\infty}^{\infty}$ is a realization of our process then $T\xi = \xi' = \{\xi_i'\}_{-\infty}^{\infty}$, where $\xi_i' = \xi_{i+1}$, i.e. ξ' is the shifted realization. In order that T be an object of study in ergodic theory it is necessary that T preserve the original probability distribution: Since $T^s\{\xi_{i_1} \in A_1, \ldots, \xi_{i_r} \in A_r\} = \{\xi_{i_1-s} \in A_1, \ldots, \xi_{i_r-s} \in A_r\}$, that means that $P\{\xi_{i_1-s} \in A_1, \ldots, \xi_{i_r-s} \in A_r\}$ must not depend on s. The processes for which the last condition is fulfilled are called strictly random processes.

Example 1. The Bernoulli automorphism.

If all ξ_i are independent then the translation T is called a Bernoulli shift or a Bernoulli automorphism. The stationarity in the strict sense reduces to the equal distribution of all the random variables ξ_i.

Example 2. The Markov automorphism.

If the ξ_i are random variables forming a finite Markov chain: $\xi_i \in E = \{e_1, e_2, \ldots, e_r\}$; $P = \| p_{ij} \|$ and $\pi = (\pi_1, \ldots, \pi_r)$ are respectively the matrix of transition probabilities and a stationary distribution, i.e. $\pi = \pi P$, then the shift T is called a Markov automorphism.

Example 3. The Gauss automorphism (normal automorphism).

If the random variables ξ_i have a joint gaussian distribution then T is called a gaussian (or normal) automorphism. As is known, the gaussian distribution is uniquely determined by its moments of the first two orders

$$a_i = \mathfrak{M}\xi_i = \int \xi_i(\omega) dP(\omega)$$

$$b_{ij} = \mathfrak{M}(\xi_i - a_i)(\xi_j - a_j) = \int (\xi_i(\omega) - a_i)(\xi_j(\omega) - a_j) dP(\omega) .$$

It follows easily from the fact that T preserves measure that

$$a_i = \text{const (i.e. } \mathfrak{M}\xi_i = \mathfrak{M}\xi_j);$$

$$b_{ij} = f(|i-j|) .$$

In other words, the stationary gaussian measure is given by two parameters: the number a and the sequence b(r). The sequence b(r) is positive definite, hence by the Bochner-Khinchin theorem the numbers b(r) can be presented as the fourier coefficients of some finite measure σ on the circle:

$$b(r) = \int_0^{2\pi} e^{i\lambda r} d\sigma(\lambda) .$$

Conversely, it is possible to construct a stationary gaussian process for every number a and every measure σ on the circle.

Now we will prove the ergodicity of the Bernoulli automorphism.

Theorem. The Bernoulli automorphism is ergodic.

Let A be an invariant mod 0 set with positive measure. This means that for all natural n $\mu(T^n A \Delta A) = 0$. Let $\mu(A) < 1$. We will carry this assumption to a contradiction.

1. The set A, like every measurable set, can be approximated by the union of finitely many finite dimensional cylinders. More accurately, for every $\varepsilon > 0$ there is a set A_ε such that $A_\varepsilon = \bigcup_{i=1}^{k} A_i$ (A_i is a cylinder depending on finitely many coordinates) and

$$\mu(A_\varepsilon \Delta A) < \varepsilon .$$

We denote by $n(\varepsilon)$ the distance from the zero coordinate to the most distant of those on which the set A_ε depends (or, which is the same, at least one of the A_i). Then the set $T^{2n(\varepsilon)+1} A_\varepsilon$ is determined by the values of the coordinates within the bounds $-3n(\varepsilon) - 1 \leq i \leq -n(\varepsilon) - 1$. By virtue of the definition of the Bernoulli measure, the sets A_ε and $T^{2n(\varepsilon)+1} A_\varepsilon$ are independent and hence

$$P(A_\varepsilon \cap T^{2n(\varepsilon)+1} A_\varepsilon) = P^2(A_\varepsilon) .$$

Furthermore we have:

$$|P(T^{2n(\varepsilon)+1}A_\varepsilon \cap A_\varepsilon) - P(A)| \le |P(T^{2n(\varepsilon)+1}A_\varepsilon \cap A_\varepsilon) - P(T^{2n(\varepsilon)+1}A \cap A_\varepsilon)| +$$

$$+ |P(T^{2n(\varepsilon)+1}A \cap A_\varepsilon) - P(T^{2n(\varepsilon)+1}A \cap A)| \le P(T^{2n(\varepsilon)+1}(A_\varepsilon \Delta A)) + P(A_\varepsilon \Delta A) \le 2\varepsilon,$$

$$|P(A) - P^2(A_\varepsilon)| \le 2\varepsilon, \quad |P^2(A) - P^2(A_\varepsilon)| \le 2\varepsilon .$$

hence $|P(A) - P^2(A)| \le 4\varepsilon$ or, in view of the arbitrariness of ε, $P(A) = P^2(A)$, i.e. $P(A) = 1, 0$, which completes the proof.

Markov automorphisms.

We will examine Markov chains with finitely many states. A stationary probability distribution for each such chain is determined by a stochastic matrix $P = \|p_{ij}\|$ and a vector of stationary probability $\pi = (\pi_1, \ldots, \pi_r)$.

We recall the well-known classification of the states of finite Markov chains (see, for example, Feller "Introduction to the theory of probability and its applications", vol. 1). Every Markov chain decomposes into classes of essential communicating states. Each such class has a certain number d of cyclic subclasses (this number is called the period of the class). In addition, the Markov chain has states from which it is possible to leave, but is not possible to return (they are called nonessential states). If there are several classes then the stationary distribution is not unique, and if the Markov chain consists of only one class then it is unique. In the stationary case the probability of a nonessential state is equal to zero.

A necessary and sufficient condition for ergodicity of a Markov automorphism is that the Markov chain consist of precisely one class.

If $d = 1$ then the Markov automorphism satisfies the mixing property, and if $d > 1$ then there is no mixing. In this way, from the point of view of ergodic theory the ideal case is that chain which consists of precisely one class and one subclass. As is known, this holds if and only if all elements of some power of the matrix of transition probabilities are strictly positive. Now we will show that a Markov automorphism may arise from problems of an entirely nonprobabilistic character.

Automorphisms of the two dimensional torus and Markov automorphisms.

Now we will speak about a construction of Adler and Weiss. Let $\mathcal{M} = T^2$ be a two dimensional torus examined as an additive abelian group. By an automorphism of the torus we will understand an automorphism of the torus as an abelian group.

In such a case, the transformation T is a linear transformation with matrix $T = \begin{vmatrix} a & b \\ c & d \end{vmatrix}$. In order that T be a transformation of the torus it is necessary that points of the plane whose coordinates differ by integers be transformed into points whose coordinates differ by integers. It follows from this that a, b, c, d must be integers. From the requirement of invariance of the measure we obtain $\det \begin{vmatrix} a & b \\ c & d \end{vmatrix} = \pm 1$. The transformation T has two eigenvalues. If they are complex then they are conjugate and each of them has modulus 1 (since their product equals ± 1). In this case it is not difficult to show that T is not ergodic. We will examine the case when λ_1 and λ_2 are real. We denote their corresponding eigenvectors by e_1, e_2 and we suppose for definiteness

$$|\lambda_1| > 1, \quad |\lambda_2| < 1 .$$

We will examine the partition of the torus into the sets

$$\xi = \{\mathcal{A}_1, \mathcal{A}_2, \ldots, \mathcal{A}_n\}$$

so that $\bigcup_{i=1}^{r} \mathcal{A}_i = \mathcal{M}$ mod 0, $\mu(\mathcal{A}_i \cap \mathcal{A}_j) = 0$ if $i \neq j$. We will construct a mapping, which is standard in ergodic theory, of $\mathcal{M}$ into the space of sequences of r symbols (Ω_r).

Let $x \in \mathcal{M}$ be an arbitrary point of the torus and let the point $T^s x \in \mathcal{A}_{i_s}$ (the s-th power of the transformation T, applied to the point x, lies in the set $\mathcal{A}_{i_s}$). We define the mapping $\varphi : \mathcal{M} \to \Omega_r$ setting

$$\varphi(x) = \{i_s, -\infty < s < \infty\} .$$

The mapping φ induces a measure on the space Ω_r: for any set C in Ω_r

$$P(C) = \mu(\varphi^{-1}(C)) .$$

Since $\varphi(Tx) = T\varphi(x)$ then, if T also denotes a shift in Ω_r, we have

$$P(TC) = \mu(\varphi^{-1}(TC)) = \mu(T\varphi^{-1}(C)) = \mu(\varphi^{-1}C)) = P(C).$$

This means that the resulting measure P is stationary in Ω_r.

The following question arises naturally: what is the set of random processes that can be obtained from the transformation T (taking different partitions ξ)? We will prove that, in the case of an automorphism of the torus, it is possible to choose a partition ξ such that the corresponding measure is markovian.

Let each of the sets $\mathcal{A}_i$ be a parallelogram whose sides are parallel to the vectors e_1 and e_2.

The sides parallel to e_1 will be called the expanding boundary of the parallelogram, and the sides parallel to e_2 will be called the contracting boundary (denoted, correspondingly, $\gamma^{(e)}(\mathcal{A}_i)$ and $\gamma^{(c)}(\mathcal{A}_i)$).

The contracting boundary of a partition ξ is the union of the contracting boundaries of all parallelograms $\mathcal{A}_i$. The expanding boundary of the partition ξ is defined analogously. (We denote them $\gamma^{(c)}(\xi)$ and $\gamma^{(e)}(\xi)$).

$T\xi$ is also a partition of $\mathcal{M}$ into sets $T\mathcal{A}_i$, which will be parallelograms of the same form.

<u>Definition.</u> The partition ξ is called markovian if

1) $\gamma^{(c)}(T\xi) \subseteq \gamma^{(c)}(\xi)$;

2) $\gamma^{(e)}(T\xi) \supseteq \gamma^{(e)}(\xi)$.

<u>Theorem.</u> If ξ is a markovian partition then the induced measure in the space of sequences is a measure corresponding to a Markov chain.

<u>Proof.</u> Let $\mathcal{A}_{i_0}$ be a parallelogram. Then the markovian condition implies that every intersection $\mathcal{A}_{i_0} \cap T^{-1}\mathcal{A}_{i_1} \cap \ldots \cap T^{-n}\mathcal{A}_{i_n}$ is, as before, a parallelogram or several parallelograms lying within $\mathcal{A}_{i_0}$. It is essential here that one pair of sides of each parallelogram inside $\mathcal{A}_{i_0}$ lies on the corresponding pair of sides of $\mathcal{A}_{i_0}$. The parallelogram $T\mathcal{A}_{i_{-1}}$ is stretched in the

other direction and $\mathcal{A}_{i_0} \cap T\mathcal{A}_{i_{-1}}$ is the union of several parallelograms, each of which has a pair of sides on another pair of opposite sides of the parallelogram $\mathcal{A}_{i_0}$. It follows from this that the conditional probability

$$\mu(T\mathcal{A}_{i_{-1}} | \mathcal{A}_{i_0} \cap T^{-1}\mathcal{A}_{i_1} \cap \ldots \cap T^{-n}\mathcal{A}_{i_n}) =$$

$$= \frac{\mu(T\mathcal{A}_{i_{-1}} \cap \mathcal{A}_{i_0} \cap T^{-1}\mathcal{A}_{i_1} \cap \ldots \cap T^{-n}\mathcal{A}_{i_n})}{\mu(\mathcal{A}_{i_0} \cap T^{-1}\mathcal{A}_{i_1} \cap \ldots \cap T^{-n}\mathcal{A}_{i_n})}$$

$$= \frac{\mu(T\mathcal{A}_{i_{-1}} \cap \mathcal{A}_{i_0})}{\mu(\mathcal{A}_{i_0})} = \mu(T\mathcal{A}_{i_{-1}} | \mathcal{A}_{i_0}) ,$$

which had to be proved.

We will introduce a new important concept.

Let $\mathcal{M}$ be a measurable space on which a measure μ is given, and let T be a measure preserving bijection of this space. We will examine the partition ξ of the space $\mathcal{M}$:

$$\xi = (C_1, C_2, \ldots, C_r) ,$$

$$\mu(C_i \cap C_s) = 0 \text{ if } i \neq j, \; \bigcup_{i=1}^{r} C_i = \mathcal{M} \bmod 0.$$

<u>Definition.</u> The partition ξ is called generating for the transformation T if the smallest closed σ-algebra containing the sets $T^k C_i$, $-\infty < k < \infty$, $i = 1, \ldots, r$ coincides with the σ-algebra of all measurable sets in $\mathcal{M}$.

Let $\mathcal{M}$ be a Lebesgue space with a continuous measure, i.e. a space that admits an isomorphism mod 0 with the interval $[0, 1]$ with Lebesgue measure. The Lebesgue space satisfies the following properties.

Let there be given a system of subsets $\{B_1, B_2, \dots\}$ of the space $\mathcal{M}$. It generates a countable dense set of the σ-algebra of measurable sets of the Lebesgue space if and only if it is possible to exclude from the Lebesgue space a set $\mathcal{N}$ of measure zero such that for any two points x and y from $\mathcal{M}-\mathcal{N}$ there exists a set B_i for which one of the following alternatives holds: either

$$x_i \in B_i,\ y_i \notin B_i \qquad \text{or} \qquad x_i \notin B_i,\ y_i \in B_i .$$

This theorem belongs to V. A. Rohlin. Its proof can be found in the paper of V. Rohlin "Fundamental concepts of measure theory" (Mat. Sbornik, vol. 25, 1949).

Let $\xi = (\mathcal{A}_1, \dots, \mathcal{A}_r)$ be a markovian partition for an automorphism of the torus T.

Lemma. If all possible intersections $G = T^{-k}\mathcal{A}_{i_k} \cap \dots \cap T^{\ell}\mathcal{A}_{i_{-\ell}}$, $k > 0$, $\ell > 0$ are connected then the partition ξ is generating for T.

Proof. It follows from the fact that ξ is a markovian partition, as was explained above, that every intersection G is a parallelogram. Again, by virtue of the markovian property $L[\gamma^{(e)}(G)] \leq \lambda_1^{-k} L[\gamma^{(e)}(\mathcal{A}_{i_k})]$, $L[\gamma^{(c)}(G)] \leq \lambda_2^{\ell} L[\gamma^{(c)}(\mathcal{A}_{i_{-\ell}})]$ where L is the length.

It follows from this that diam G tends to zero when k, ℓ tend to infinity. Let

$$N = \bigcup_{i=1}^{r} \bigcup_{n=-\infty}^{\infty} (T^n(\gamma^{(e)}(\mathcal{A}_i) \cup \gamma^{(c)}(\mathcal{A}_i))) .$$

Then, if x, y belong to $\mathcal{M}-\mathcal{N}$, there exists G such that

$$\operatorname{diam}(T^{-k}\mathcal{A}_{i_{-k}} \cap \ldots \cap T^{k}\mathcal{A}_{i_k}) < d(x, y), \quad \text{where } x \in T^{-k}\mathcal{A}_{i_{-k}} \cap \ldots \cap T^{k}\mathcal{A}_{i_k} .$$

The lemma has been proved. Now we will construct one of the markovian partitions.

Let T be an automorphism of the two dimensional torus $\mathfrak{M}$. It always has a fixed point, the point 0. Let e_1 and e_2 be the expanding and contracting eigenvectors of the transformation T, and $\gamma^{(1)}$, $\gamma^{(2)}$ be infinite lines on the torus passing through e_1 and e_2, respectively. Every interval $\ell^{(1)}$ of the line $\gamma^{(1)}$ containing the point 0 satisfies the property that $T\ell^{(1)} \supset \ell^{(1)}$. Analogously, for an interval $\ell^{(2)}$ on the line $\gamma^{(2)}$ containing the point 0 we have $T\ell^{(2)} \subset \ell^{(2)}$.

Now we will construct a partition ξ of the torus $\mathfrak{M}$ into two parallelograms such that the boundary $\gamma^{(e)}(\xi)$ will be an interval $\ell^{(1)} \in \gamma^{(1)}$ containing the point 0 and, analogously, $\gamma^{(c)}(\xi)$ will be an interval of the line $\gamma^{(2)}$ containing the point 0. By virtue of the above, every partition ξ into such sets will be markovian.

We take an arbitrary interval a_1 coming out of the point 0 in the direction $\ell^{(1)}$ (fig. 1). We draw the interval a_2 from the point 0 in the direction e_2 until the intersection with a_1 (fig. 2). Now we extend a_1 as far as the intersection with a_2 and we denote that interval a_3 (fig. 3). We extend a_3 in the opposite direction from the point 0 as far as its intersection with a_2. We denote the entire resulting interval, lying on $\gamma^{(1)}$, by a_4 (fig. 4)

Fig. 1

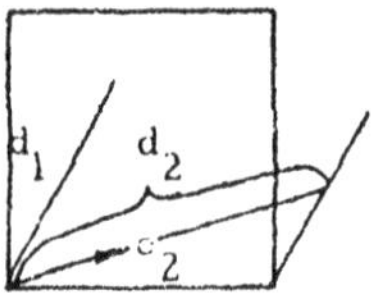

Fig. 2

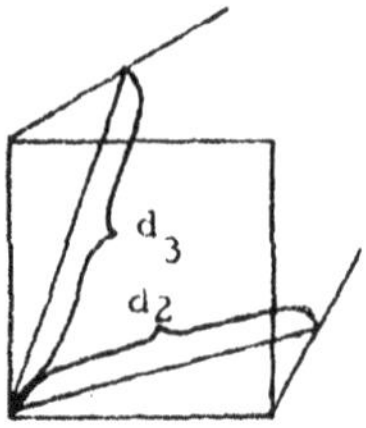

Fig. 3

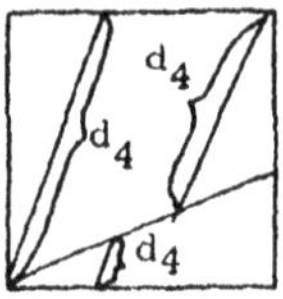

Fig. 4

One of the parallelograms of the resulting partition is cross-hatched on fig. 5.

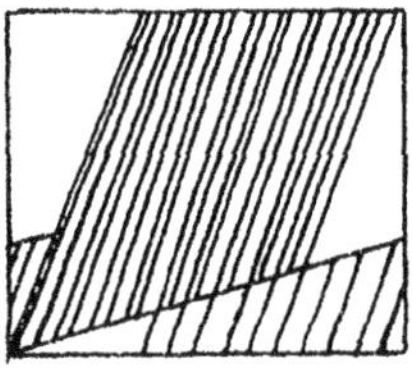

Fig. 5

Generally speaking, the markovian partition ξ that we constructed above may not be generating (and, as a rule, it will not be). However, a generating markovian partition can be easily obtained in the following way.

We will examine the partition ξ_1 whose elements have the form $\mathcal{A}_i \cap T\mathcal{A}_j$. The elements of this partition may not be connected. We introduce the partition η whose elements are the connected components of the partition ξ_{ℓ}. The partition η is again markovian, and all corresponding intersections are connected. Hence the partition η is generating for T.

In abstract ergodic theory the concept of isomorphism of two measure preserving transformations is introduced, namely: Let T_1 be a measure preserving bijection (mod 0) of the space $\mathcal{M}_1$, and let T_2 be a measure preserving bijection (mod 0) of the space $\mathcal{M}_2$. Then T_1 and T_2 are said to be isomorphic (mod 0) if there exists an isomorphism (mod 0) $\mathcal{U}$ from the space $\mathcal{M}_1$ onto the space $\mathfrak{M}_2$ such that

$$\mathcal{U}T_1 = T_2\mathcal{U} \text{ (mod 0)}, \; T_1\mathcal{U}^{-1} = \mathcal{U}^{-1}T_2 \text{ (mod 0)}$$

The term mod 0 always means that equality becomes exact if some sets of measure zero are excluded from the corresponding spaces.

The assertion proved above can now be formulated in the following way: An automorphism of the torus is isomorphic (mod 0) to a Markov automorphism.

References

1. R. Adler and B. Weiss, Similarity of automorphism of the torus, Memoirs Amer. Math. Soc. 1970, vol. 98, 1-43.

2. The concept of a Markov Partition was introduced in Ja. G. Sinai, Markov Partitions and C-diffeomorphisms" Funct. Analysis and Appl. v. 2 1(1968), 64-81.

More general and more useful definition is in R. Bowen, Markov Partitions and minimal sets for axiom a diffeomorphisms" Amer. J. of Math., 92:4(1970), 907-918

Lecture 13

GAUSSIAN SYSTEMS

Let Ω be the space of sequences $\omega = \{\omega_i\}_{-\infty}^{\infty}$, infinite in both directions, where each coordinate $\omega_i \in \mathbb{R}^1$.

The space of such sequences is met in the theory of probability as the space of elementary events of a real valued random process.

We suppose that a probability measure μ is given on the space Ω. Such a measure is called gaussian if every finite subset of the set of random variables $(\ldots\xi_{-n},\ldots,\xi_0,\ldots,\xi_n\ldots)$ has a multivariate gaussian distribution as joint distribution.

As is known, every gaussian distribution is given by its first two moments: and function of one variable m_i, where $m_i = \mathfrak{M}\xi_i$, and a function of two variables $\mathfrak{P}(s,t)$, where $\mathfrak{P}(s,t) = \mathfrak{M}(\xi_s - m_s)(\xi_t - m_t)$.

We will examine the shift transformation T on Ω:

$$T\omega = \omega', \ \omega_i' = \omega_{i+1} .$$

It is clear that this transformation is measurable, i.e. it transforms a measurable set into a measurable set. In the case of a gaussian measure it is not difficult to check that for the invariance of the measure μ it is necessary and sufficient that $m_i = m = \text{const}$ and $\mathfrak{P}(s,t) = \mathfrak{P}(s-t)$. In this case the function of one variable $\mathfrak{P}(s)$ is a positive definite function and, in this way, by the Bochner-Khinchin theorem, $\mathfrak{P}(s) = \int_0^{2\pi} e^{i\lambda s} d\sigma(\lambda)$, where σ is a finite measure on S^1 that is invariant with respect to the transformation $z \to \bar{z}$. The measure

σ is called the spectral measure of the gaussian process.

Definition. The transformation T with the invariant gaussian measure μ is called a gaussian automorphism.

Since the measure μ is uniquely determined by m and the measure σ, it is clear that all the ergodic properties of a gaussian automorphism can be expressed through these parameters. These properties should not depend on m since the translation $\Omega^{\pi} \to \Omega$, $\pi(\omega) = \{\omega_i - m\}$ of the numbering origin of the real numbers transforms the gaussian measure with parameters m, σ into the gaussian measure $0, \sigma$.

In the future we will always take $m = 0$.

Theorem (Fomin, Maruyama).

I. The transformation T is ergodic if and only if the measure σ is nonatomic.

II. The transformation T satisfies the mixing property if and only if $\mathfrak{G}(s) \to 0$ as $s \to \infty$.

Consequence. If the measure σ is absolutely continuous with respect to Lebesgue measure, i.e. $d\sigma(\lambda) = r(\lambda)\, d\lambda$, then T satisfies the mixing property. This follows from the Riemann-Lebesgue theorem, according to which the Fourier coefficients of an integrable function converge to zero.

Proof of I.

It is easy to establish the necessity of our condition.

We will examine in $\mathcal{L}^2(\Omega,\mu)$ the sequence of functions $f_i(\omega) = \omega_i$, $-\infty < i < \infty$. It is clear that $f_i(\omega) = \mathcal{U}^i f_0$, where $\mathcal{U}$ is the unitary operator generated by the gaussian shift T.

We construct the linear space H_1 consisting of the closure of all possible linear combinations $\sum_s c_s f_s$, where $c_s \in \mathbb{C}$. It is clear that $\mathcal{U}H_1 = H_1$ and every random variable $h \in H_1$ has a gaussian distribution.

Now we will obtain a more convenient description for the subspace H_1.

We assign to every linear combination $\Sigma c_s fs$ the trigonometric polynomial $\Sigma c_s e^{i\lambda s} = P(\lambda)$. If we introduce into the space of polynomials $P(\lambda)$ a scalar product according to the formula $(P_1, P_2) = \int_0^{2\pi} P_1\bar{P}_2 d\sigma(\lambda)$, then the correspondence that we constructed will be a linear isometric transformation, and hence can be extended by continuity to the corresponding closure. The closure of the linear combinations $\Sigma c_s fs$ is the space H_1, and the closure of the trigonometric polynomials is $\mathcal{L}^2(S^1, \sigma)$.

We will denote the transformation that we constructed from H_1 into $\mathcal{L}^2(S^1, \sigma)$ by α.

The fundamental property of the transformation α follows from the relation $\alpha\mathcal{U} = e^{i\lambda}\alpha$ or, in more detail,

$$\alpha(\mathcal{U}h) = \alpha(\mathcal{U}\sum_s c_s fs) = \alpha(\Sigma c_s f_{s+1}) = e^{i\lambda}\alpha(h)$$

for every $h \in H_1$.

Let there be a point λ_0 for which $\sigma(\lambda_0) > 0$.

We will examine an element of the Hilbert space H_1 corresponding to the function $\varphi(\lambda) = \begin{cases} 1 & \lambda = \lambda_0 \\ 0 & \lambda \neq \lambda_0 \end{cases}$. Since

$\int |\varphi(\lambda)|^2 d\sigma(\lambda) = \sigma(\lambda_0) > 0$, we have $\|h\|^2 = \sigma(\lambda_0)$. Since $e^{i\lambda}\varphi(\lambda) = e^{i\lambda_0}\varphi(\lambda)$, we have $\mathcal{U}h = e^{i\lambda_0}h$.

It follows from this that $|h(\omega)|$ is an invariant function. Indeed,

$$\mathcal{U}|h(\omega)| = |h(T\omega)| = |e^{i\lambda_0}h(\omega)| = |h(\omega)| .$$

But $\|h(\omega)\| > 0$ and h is a gaussian random variable distribution. Hence $|h(\omega)|$ is not constant and, consequently, T is not ergodic.

The proof of sufficiency in I is based on the important theory of Ito of multiple stochastic integrals.

One of the results of this theory states that the Hilbert space $\mathcal{L}^2(\Omega,\mu)$ can be decomposed into the orthogonal direct sum of subspaces $H_{n,m}$, where each space $H_{n,m}$ is isomorphic to the space of $(n+m)$-dimensional functions $\varphi(\lambda_1,\ldots,\lambda_n;\lambda_1',\ldots,\lambda_m')$ defined on the $(n+m)$-dimensional torus and square integrable with respect to the measure $\underbrace{\sigma \times \sigma \times \ldots \times \sigma}_{n+m}$, even in each variable and symmetric on each of the groups of variables $\lambda_1,\ldots,\lambda_n;\lambda_1',\ldots,\lambda_m'$ separately. It is most essential that the operator $\mathcal{U}$ is transformed into the operator of multiplication by $\exp[i(\lambda_1+\ldots+\lambda_n-\lambda_1'-\ldots\lambda_m')]$ under this isomorphism. Also, each subspace $H_{n,m}$ is invariant with respect to $\mathcal{U}$.

We will prove the sufficiency in I using these facts.

Let T be nonergodic, i.e. there exists an invariant nonconstant function h. One of its projections on a subspace $H_{m,n}$ $(m+n\ 0)$ must be different from zero. Each projection $h_{n,m}$ is again an invariant function (Prove it!).

Let $\varphi_{n,m}$ correspond to $h_{n,m}$ under the above described isomorphism. The invariance condition on $h_{n,m}$ means that $e^{i(\lambda_1+\ldots+\lambda_n-\lambda'_1-\ldots-\lambda'_m)}\varphi_{n,m} =$ $= \varphi_{n,m}$ holds almost everywhere with respect to the measure $\underbrace{\sigma\times\ldots\times\sigma}_{n+m}$.

It follows from this that $\varphi_{n,m} \neq 0$ only on the hyperplane $\lambda_1+\ldots+\lambda_n-\lambda'_1-\ldots-\lambda'_m = 0$. But since the measure σ is continuous, the measure of that hyperplane equals 0. In this way, $\varphi_{n,m} = 0$ almost everywhere for all (n, m).

Assertion I has been proved completely.

Before proving assertion II we will describe the Ito construction (for a complex gaussian system).

We write the spectral representation for a stationary gaussian sequence

$$x(t) = \int_0^{2\pi} e^{i\lambda t} dF(\lambda) .$$

For each Borel subset Λ of the interval $[0, 2\pi]$ the variable $F(\Lambda)$ is a gaussian random variable with mean 0 and variance $\sigma(\Lambda)$. Furthermore, for nonintersecting sets Λ_1 and Λ_2 the random variables $F(\Lambda_1)$ and $F(\Lambda_2)$ are mutually independent.

The linear space H_1 can be conceived as the space of one-dimensional stochastic integrals of the form

$$\int_0^{2\pi} \varphi(\lambda)\, dF(\lambda) .$$

The natural generalization of the one-dimensional stochastic integral must be the multidimensional stochastic integral of the form

$$\int \varphi(\lambda_1 \ldots \lambda_n ; \lambda'_1 \ldots \lambda'_m) dF(\lambda_1) \ldots dF(\lambda_n) \overline{dF(\lambda'_1)} \ldots \overline{dF(\lambda'_m)} .$$

However, the correct definition of such an integral must be introduced with some caution. If φ_1 and φ_2 differ only in a permutation of the arguments λ or λ' then the result of the integration must yield one and the same random variable. It follows from this that, already from the very beginning, it is necessary to examine only functions which are symmetric with respect to λ and with respect to λ'.

Another, more substantial circumstance, relates to the stochastic character of the random measure F.

The following nonrigorous reasoning should help the reader clarify the essence of the situation.

We will examine the expression $dF(\lambda_1)\dots dF(\lambda_k)\cdot\overline{dF(\lambda'_1)}\dots\overline{dF(\lambda'_\ell)}$ and we will suppose that the transformation $\mathcal{U}$ can be applied to it. Since $\mathcal{U}dF(\lambda_k) = e^{i\lambda_k}dF(\lambda_k)$, $\mathcal{U}\overline{dF(\lambda'_\ell)} = e^{-i\lambda'_\ell}\overline{dF(\lambda'_\ell)}$ and the operator $\mathcal{U}$ is multiplicative, we have

$$\mathcal{U}(dF(\lambda_1)\dots dF(\lambda_k)\overline{dF(\lambda'_1)}\dots\overline{dF(\lambda'_\ell)}) =$$

$$= e^{i(\sum_{s=1}^{k}\lambda_r - \sum_{s=1}^{\ell}\lambda'_s)} dF(\lambda_1)\cdot\ldots\cdot dF(\lambda_k)\cdot\overline{dF(\lambda'_1)}\dots\overline{dF(\lambda'_\ell)}.$$

If $\sum_{0=1}^{k}\lambda_s - \sum_{s=1}^{\ell}\lambda'_s = 0$ then that means that

$$dF(\lambda_1)\dots dF(\lambda_k)\overline{dF(\lambda'_1)}\dots\overline{dF(\lambda'_\ell)}$$

is an invariant function of the operator $\mathcal{U}$. In the ergodic case every invariant function equals its mathematical expectation. In the gaussian case this mathematical expectation is different from zero only in the case $k = \ell$ and is concentrated on the subspaces $\lambda_1 = \lambda'_{i_1}, \lambda_2 = \lambda'_{i_2}, \ldots, \lambda_k = \lambda'_{i_k}$. In this case one must set

$$dF(\lambda_1)\cdot\ldots\cdot dF(\lambda_k)\overline{dF(\lambda'_1)}\cdot\ldots\cdot\overline{dF(\lambda'_k)} = \prod_{i=1}^{k} |dF(\lambda_i)|^2 = \prod_{i=1}^{k} d\sigma(\lambda_i) .$$

Hence, in the definition of this stochastic integral, it is necessary to exclude from the region of integration all possible subspaces $\lambda_i = \lambda'_j$. A detailed construction can be found in the paper of Ito:

"Complex Multiple Wiener Integral", Japan J. Math., 22, 63-86, 1949.

In addition to the correct construction of multidimensional stochastic integrals, this paper contains a proof of the fact that linear combinations of multiple stochastic integrals form a dense set in $\mathcal{L}^2(\Omega, \mu)$.

Each space $H_{n,m}$ is a space of stochastic integrals of the type

$$\int \varphi(\lambda_1, \ldots, \lambda_n; \lambda'_1, \ldots, \lambda'_m) dF(\lambda_1)\ldots dF(\lambda_m)\overline{dF(\lambda'_1)}\ldots\overline{dF(\lambda'_m)} .$$

We have described the Ito construction for a complex gaussian system. For a real gaussian system, in order to construct a stochastic integral it is necessary to contract the collection of functions, using only functions $\varphi(\lambda_1, \ldots, \lambda_n)$ satisfying the relation

$$\varphi(\lambda_1, \ldots, \lambda_{i-1}, -\lambda_i, \lambda_{i+1}, \ldots, \lambda_n) = \overline{\varphi(\lambda_1, \ldots, \lambda_n)}$$

for all i, $1 \le i \le n$.

We turn to the proof of condition II.

As before, for the proof of mixing it is sufficient to show that $(\mathcal{U}^k h, h) \to 0$ as $k \to \infty$ for all h such that $\mathfrak{M} h = (h, 1) = 0$. Therefore, if there is mixing, then for the function $h = \omega_0$

$$\mathfrak{C}(s) = \mathfrak{M}\omega_{st} \cdot \omega_0 \to \mathfrak{M}\omega_0 \cdot \mathfrak{M}\omega_0 = 0,$$

i.e., our condition is necessary. In order to prove sufficiency we will examine the projection of h on the subspace $H_{n,m}$.

Since

$$(\mathcal{U}^s h_{n,m}, h_{n,m}) = \int_0^{2\pi} e^{is(\Sigma\lambda_i - \Sigma\lambda_j')} |\varphi(\lambda_1 \dots \lambda_n; \lambda_1' \dots \lambda_m')|^2$$

$$\cdot\, d\sigma(\lambda_1) \dots d\sigma(\lambda_n) d\sigma(\lambda_1') \dots d\sigma(\lambda_m') .$$

and $\mathfrak{C}(s) = \int_0^{2\pi} e^{is\lambda} d\sigma(\lambda) \to 0$ as $s \to \infty$ by hypothesis, by virtue of the generalized Riemann-Lebesgue theorem $(\mathcal{U}^s h_{n,m}, h_{n,m}) \to 0$ as $s \to \infty$. This assertion is true for arbitrary $h_{n,m}$.

Assertion II has been proved completely.

Here are some more details. We can rewrite the last expression in the following way

$$(\mathcal{U}^s h_{n,m}, h_{n,m}) = \int_0^{2\pi} e^{is\mu} d\sigma_{n,m}(\mu | \varphi),$$

where the measure $\sigma_{n,m}(\cdot | \varphi)$ is absolutely continuous under the measure $\underbrace{\sigma * \sigma * \dots * \sigma}_{(n+m)\cdot \text{times}}$ (here $*$ stands for convolution) and is defined in the equality:

$$\sigma_{n,m}(A|\varphi) = \int_{\Sigma\lambda_i - \Sigma\lambda_j' \in A} |\varphi(\lambda_1 \dots \lambda_n; \lambda_1', \dots, \lambda_m')|^2 d\sigma(\lambda_1) \cdot \dots \cdot d\sigma(\lambda_n) d\sigma(\lambda_1') \dots d\sigma(\lambda_m') .$$

The fourier coefficients of the convolution $\underbrace{\sigma * \dots * \sigma}_{n+m}$ tend to zero. The generalized

Riemann-Lebesgue theorem asserts that if σ is an arbitrary measure the fourier coefficients of which tend to zero and σ' is absolutely continuous under σ then fourier coefficients of σ' also tend to zero.

References

1. S. V. Fomin, Normal dynamical systems, Ukranian Math. J., 1950, v 2, n2.

2. G. Maruyame,

3. I. V. Ojirsanov, On spectre of dynamical systems generated by stationary random processes. Doklady, 1959, v. 126, n5, 931-934.

4. A. M. Vershik, Concerning the theory of normal dynamical systems. Doklady, 1962, vol. 144, nl, -12; On spectral and metric isomorphism of some normal dynamical systems, Doklady, 1962, v. 144, n2, 255-257.

Lecture 14.

THE ENTROPY OF A DYNAMICAL SYSTEM

To each dynamical system corresponds a certain number (possibly infinite) called its entropy. The concept of entropy was introduced into ergodic theory by A. N. Kolmogoroff in 1959. It is most interesting that, as often happens in these cases, when the entropy is positive the dynamical systems satisfy a whole series of strong additional statistical properties.

The word "entropy" is used in many senses in mathematics and physics. In equilibrium statistical physics entropy is understood to mean the coefficient of the asymptotics of the logarithm of the number of configurations satisfying these or those properties when the number of degrees of freedom tends to infinity. If this concept of entropy is used then the entropy met in ergodic theory can be called dynamical in a natural way: It is the coefficient of the asymptotics of the logarithm of the number of different types of trajectories of the dynamical system when the time tends to infinity.

In order to give a more exact definition, it is necessary to begin with the entropy of partitions and its simplest properties.

We will denote by $(\mathcal{M}, \gamma, \mu, T)$ a dynamical system, $\alpha = \{\mathcal{A}_i\}$, $i \in I$, $\beta = \{B_j\}$, $j \in \mathcal{J}$ will be a finite or countable partition of $\mathcal{M}$. We set

$$z(x) = \begin{cases} -x \ln x & \text{if} \quad 0 < x \leq 1 \\ 0 & \text{if} \quad x = 0 \end{cases}$$

We always understand ln to be the natural logarithm; $z(x)$ is a continuous and concave function on $[0,1]$.

Definition 1. The entropy $H(\alpha)$ of the partition α is

$$H(\alpha) = \sum_{i \in I} z(\mu(\mathcal{A}_i)).$$

Exercise. If card $I = N < \infty$ then $H(\alpha) \leq \ln N$. $H(\alpha) = \ln N$ if and only if $\mu(\ _i) = \frac{1}{N}$ for all i. $H(\alpha) = 0$ if and only if $N = 1$.

Definition 2. The conditional entropy of α with respect to β, $H(\alpha|\beta)$, is

$$H(\alpha|\beta) = \sum_{j \in J} \mu(B_j)[\sum_{i \in I} z(\mu(\mathcal{A}_i|B_j))] .$$

Theorem 1. If $\alpha' = \{\mathcal{A}'_k\}_{k \in K}$, $\beta' = \{B'_n\}_{n \in N}$ are measurable partitions of $\mathcal{M}$ then

a) $H(\alpha \vee \beta|\beta) = H(\alpha|\beta)$;

b) $H(\alpha|\beta) = 0$ if and only if $\alpha \leq \beta$;

c) $H(\alpha \vee \beta) = H(\beta) + H(\alpha|\beta)$;

d) $\alpha \leq \alpha' \Rightarrow H(\alpha) \leq H(\alpha')$;

e) $\alpha \leq \alpha' \Rightarrow H(\alpha|\beta) \leq H(\alpha'|\beta)$;

f) $H(\alpha \vee \alpha'|\beta) \leq H(\alpha|\beta) + H(\alpha'|\beta)$;

g) $H(\alpha \vee \alpha' \vee \beta) + H(\beta) \leq H(\alpha \vee \beta) + H(\alpha' \vee \beta)$;

h) $\beta \leq \beta' \Rightarrow H(\alpha|\beta') \leq H(\alpha|\beta)$;

i) $H(\alpha \vee \beta|\beta') = H(\beta|\beta') + H(\alpha|\beta \vee \beta')$.

The proof is left to the reader as an exercise.

For any automorphism T and any partition $\alpha = \{\mathcal{A}_i\}$ we denote by $T\alpha$ the partition into the sets $\{T\mathcal{A}_i\}$. We observe that $H(\alpha|\beta) = H(T\alpha|T\beta)$ always holds.

Definition 3. The entropy per unit of time of the partition α is the number

$$h(\alpha, T) = \lim_{n\to\infty} \frac{1}{n} H(\alpha \vee T\alpha \vee \ldots \vee T^n\alpha) = \lim_{n\to\infty} H(T^n\alpha | \alpha \vee T\alpha \vee \ldots \vee T^{n-1}\alpha) .$$

The existence of this limit follows from part c) of Theorem 1. (The sequence defining $h(\alpha, T)$ consists of positive elements.)

Definition 4. The entropy of $(\mathfrak{M}, \gamma, \mu, T)$ is

$$h(T) = \sup h(\alpha, T)$$

where the supremum is taken over all finite or countable measurable partitions α of the space $\mathfrak{M}$, for which $H(\alpha) < \infty$.

Theorem 2. The entropy is an invariant of the dynamical system. In other words, isomorphic dynamical systems have the same entropy.

Proof. Let $\mathcal{U} : \mathfrak{M} \to \mathfrak{M}'$ be an isomorphism of the dynamical systems $(\mathfrak{M}, \gamma, \mu, T)$ and $(\mathfrak{M}', \gamma', \mu', T')$, i.e. $T' = \mathcal{U} T \mathcal{U}^{-1}$. For every partition α we have

$$h(\mathcal{U}\alpha, T') = h(\mathcal{U}\alpha, \mathcal{U}T\mathcal{U}^{-1}) = \lim_{n\to\infty} \frac{1}{n} H(\mathcal{U}\alpha \vee \ldots \vee \mathcal{U}T^n\alpha) =$$

$$= \lim_{n\to\infty} \frac{1}{n} H(\alpha \vee \ldots \vee T^n\alpha) = h(\alpha, T) .$$

This proves the theorem.

It is almost impossible to compute the entropy starting from the definition. The principal method that allows us to compute it is the theorem on generating partitions for entropy, due to A. N. Kolmogoroff. Before proceeding to it, we will prove some lemmas.

We denote by Z the set of all measurable partitions (of the space $\mathcal{M}$) with finite entropy. For $\alpha, \beta \in Z$ we set $|\alpha, \beta| = H(\alpha|\beta) + H(\beta|\alpha)$.

Lemma 1. $(Z, |\cdot, \cdot|)$ is a metric space.

Proof. $|\cdot, \cdot|$ separates points by part b) of Theorem 1. It is obvious that $|\cdot, \cdot|$ is symmetric in its arguments. The triangle inequality:

$$H(\alpha|\gamma) = H(\alpha \vee \gamma) - H(\gamma) \leq H(\alpha \vee \beta \vee \gamma) - H(\beta \vee \gamma)$$

$$+ H(\beta \vee \gamma) - H(\gamma) = H(\alpha|\beta \vee \gamma) + H(\beta|\gamma) \leq H(\alpha|\beta) + H(\beta|\gamma) .$$

Analogously, $H(\gamma|\alpha) \leq H(\gamma|\beta) + H(\beta|\alpha)$. The triangle inequality follows from this.

Lemma 2. For fixed T the function $h(\alpha, T)$ is continuous over Z, more precisely

$$|h(\alpha, T) - h(\beta, T)| \leq |\alpha, \beta|, \alpha, \beta \in Z.$$

Let $\alpha_n = \alpha \vee T\alpha \vee \ldots \vee T^n\alpha$, $\beta_n = \beta \vee T\beta \vee \ldots \vee T^n\beta$. It is easy to see that $|H(\alpha_n) - H(\beta_n)| \leq |\alpha_n, \beta_n|$. But

$$H(\alpha_n|\beta_n) \le H(\alpha|\beta_n) + H(T\alpha|\beta_n) + \ldots + H(T^n\alpha|\beta_n)$$

$$\le H(\alpha|\beta) + H(T\alpha|T\beta) + \ldots + H(T^n\alpha|T^n\beta) = (n+1)H(\alpha|\beta)$$

(since $\beta, T\beta, \ldots, T^n\beta \le \beta_n$). And so, $|\alpha_n, \beta_n| \le (n+1)|\alpha, \beta|$. This yields the required inequality.

Lemma 3. The set of all finite partitions of the space $\mathcal{M}$ is a dense subset of Z.

Proof. Let $\alpha = \{\mathcal{A}_1, \mathcal{A}_2, \ldots, \mathcal{A}_k, \ldots\}$. We set $E_n = \mathcal{M} - \bigcup_{i=1}^{n} \mathcal{A}_i$, $\alpha_n = \{\mathcal{A}_1, \mathcal{A}_2, \ldots, \mathcal{A}_n, E_n\}$. It is easy to see that $|\alpha, \alpha_n| \underset{n\to\infty}{\to} 0$.

Lemmas 2 and 3 yield the possibility of determining the entropy with the equality

$$h(T) = \sup h(\alpha, T) ,$$

where the supremum is taken over finite measurable partitions of the space $\mathcal{M}$.

If a partition α is given then we denote by $\gamma(\alpha)$ the σ-algebra generated by α. If a family γ_i, $i \in I$ of σ-algebras is given then we denote by $\bigvee_{i\in I} \gamma_i$ the σ-algebra generated by $\{\gamma_i\}$, $i \in I$.

Lemma 4. Let $\alpha_1 \le \alpha_2 \le \ldots$ be measurable partitions of $\mathcal{M}$ such that $\gamma = \bigvee_n \gamma(\alpha_n)$. Then $Z' = \{\beta : \beta$ is a finite partition and $\beta \le \alpha_n$ for some $n\}$ is dense in Z.

Proof. By Lemma 3 we must prove that, having chosen a finite partition

α and a number $\delta > 0$, we can find an n and a finite partition $\beta \leq \alpha_n$ such that $|\alpha, \beta| < \delta$. Let $\alpha = \{\mathcal{A}_1, \dots, \mathcal{A}_m\}$. By the hypothesis of the lemma, an n and a set $\mathcal{A}'_1, \dots, \mathcal{A}'_{m-1} \in \gamma(\alpha_n)$ can be found such that $\mu(\mathcal{A}_i \Delta \mathcal{A}'_i) < \varepsilon$ for $1 \leq i < m$. We define the partition β as follows: $\beta = \{B_1, \dots, B_m\}$, where $B_1 = \mathcal{A}'_1$, $B_i = \mathcal{A}'_i \backslash (\mathcal{A}'_1 \cup \dots \cup \mathcal{A}'_{i-1})$ for $1 \leq i < m$, $B_m = \mathcal{M} - \bigcup_{i<m} \mathcal{A}'_i$. It is is obvious that $\beta \leq \alpha_n$. We have

$$|\alpha, \beta| = \sum_{i=1}^{m} \mu(\mathcal{A}_i)\left[\sum_{j=1}^{m} z\left(\frac{\mu(\mathcal{A}_i \cap B_j)}{\mu(\mathcal{A}_i)}\right)\right] + \sum_{i=1}^{m} \mu(B_i)\left[\sum_{j=1}^{m} z\left(\frac{\mu(\mathcal{A}_j \cap B_i)}{\mu(B_i)}\right)\right]$$

$$= \sum_{i=1}^{m} z(\mu(\mathcal{A}_i)) - 2 \sum_{i,j=1}^{m} z(\mu(\mathcal{A}_i \cap B_j)) + \sum_{j=1}^{m} z(\mu(B_j))$$

$$\leq \sum_{i=1}^{m} z(\mu(\mathcal{A}_i)) - z \sum_{i=1}^{m} z(\mu(\mathcal{A}_i \cap B_i))$$

$$+ \sum_{j=1}^{m} z(\mu(B_j)) \leq \sum_{i=1}^{m} [z(\mu(\mathcal{A}_i)) - z(\mu(\mathcal{A}_i \cap B_i))]$$

$$+ \sum_{i=1}^{m} [z(\mu(B_i)) - z(\mu(\mathcal{A}_i \cap B_i))] .$$

But $0 \leq \mu(\mathcal{A}_i) - \mu(\mathcal{A}_i \cap B_i) < \varepsilon$, $0 \leq \mu(B_1) - \mu(\mathcal{A}_i \cap B_i) \leq \varepsilon$. If ε is sufficiently small then

$$z(\mu(\mathcal{A}_i)) - z(\mu(\mathcal{A}_i \cap B_i)) < \frac{\delta}{2m}, \quad z(\mu(B_i)) - z(\mu(\mathcal{A}_i \cap B_i)) < \frac{\delta}{2m},$$ with which

the proof is concluded.

Definition 5. A finite measurable partition α of the space $\mathcal{M}$ is called generating for T if

$$V_{-\infty}^{\infty} T^n \alpha = \varepsilon \quad (\text{i.e. } V_{-\infty}^{\infty} \gamma(T^n \alpha) = \gamma).$$

Where ε is the partition of $\mathcal{M}$ into points and equality is understood modulo zero

Theorem 3 (Of A. N. Kolmogoroff on generating partitions). If α is generating for T and $H(\alpha) < \infty$ then $h(T) = h(\alpha, T)$.

Proof. For a given partition ξ we define $\widetilde{\xi}_n = \bigvee_{k=-n}^{n} T^k \xi$. It is obvious that

$$\frac{1}{n} H(\widetilde{\xi}_n) = \frac{1}{2n} H(\xi \vee T\xi \vee \ldots \vee T^{2n}\xi) \cdot \frac{2n}{n} \to 2h(\xi, T) \qquad (*)$$

Clearly, $\widetilde{\alpha}_1 \leq \widetilde{\alpha}_2 \leq \ldots$ and $\vee \widetilde{\alpha}_n = \varepsilon$. Then, by Lemma 4, it suffices to prove that $\beta \leq \widetilde{\alpha}_n$ implies $h(\beta, T) \leq h(\alpha, T)$.

But if $\beta \leq \widetilde{\alpha}_n$ then

$$\widetilde{\beta}_m \leq (\widetilde{\alpha}_n)_m = \widetilde{\alpha}_{n+m}$$

for all m. Hence $\frac{1}{m} H(\widetilde{\beta}_m) \leq \frac{1}{n+m} H(\alpha_{n+m}) \cdot \frac{n+m}{m}$. If we increase m to infinity in the inequality then from inequality $(*)$ we obtain

$$2h(\beta, T) \leq 2h(\alpha, T).$$

The Kolmogoroff theorem has been proved.

Theorem 4. $h(T^s) = |s| \cdot h(T)$ for all $s \in \mathbb{Z}$.

Proof. It is easy to see that $h(T^s) = h(T^{-s})$. Hence we may assume that $s > 0$. Let ξ be a finite measurable partition, $\xi_1 = \widetilde{\xi}_s$. Then

$$\frac{1}{n} H(\xi_1 \vee T^s \xi_1 \vee \ldots \vee T^{sn} \xi_1) = \frac{1}{n} H(\xi \vee T\xi \vee T^2\xi \vee \ldots \vee T^{ns}\xi) \to sh(\xi, T).$$

This means $h(T^s) \geq sh(T)$. On the other hand, since $\xi \leq \xi_1$, $h(\xi, T^s) \leq$ $\leq h(\xi_1, T^s) = sh(\xi, T)$ holds. Hence $h(T^s) \leq sh(T)$.

Exercise. If there exists a generating α such that $\bigvee_{n=0}^{\infty} \alpha = \varepsilon$, $H(\alpha) < \infty$, then $h(T) = 0$.

Exercise. Prove that the entropy of the Bernoulli automorphism with probabilities $\{p_1, \ldots, p_r\}$ is equal to $\sum_{i=1}^{r} z(p_i)$.

It follows from the last assertion that there exists a continuum of non-isomorphic Bernoulli automorphisms. Recently the American mathematician D. Ornstein showed that Bernoulli automorphisms with equal entropies are isomorphic.

Exercise. Let $(\mathfrak{M}, \gamma, \mu, T)$ and $(\mathfrak{M}_1, \gamma_1, \mu_1, T_1)$ be dynamical systems. Prove that

$$h(T \times S) = h(T) + h(S),$$

where $(T_1 \times T_2)(x, y) = (T_1 x, T_2 y)$ is an automorphism of the measurable space $(\mathfrak{M}_1 \times \mathfrak{M}_2, \gamma_1 \times \gamma_2, \mu_1 \times \mu_2)$.

Now we will generalize Theorem 4 to the case of continuous time.

Definition 6. We will call the flow $(\mathfrak{M}, \gamma, \mu, \{S_t\})$ continuous if $\lim_{t \to 0} \mu(S_t \mathcal{A} \Delta \mathcal{A}) = 0$ for every set $\mathcal{A} \in \gamma$.

Theorem 5. If $(\mathfrak{M}, \gamma, \mu, \{S_t\})$ is a continuous flow then $h(S_t) = |t| h(S_1)$.

Proof (by M. S. Pinsker). Since $h(T) = h(T^{-1})$, we can set $t > 0$. We will prove that $0 < \mu < t$ implies $h(S_t) \le \frac{t}{\mu} h(S_\mu)$. Let m be a natural number. We set $\delta = \frac{1}{m}$. Having chosen a partition ξ, we set

$$\eta = \xi \vee S_{\delta\mu}\xi \vee S_{2\delta\mu}\xi \vee \dots \vee S_{(m-1)\delta\mu}\xi .$$

By $k = k(n)$ we denote a natural number such that $nt \le k\mu < (n+1)t$, and by $r(p)$ one such that $r(p) \cdot \delta \cdot \mu \le pt < (r(p)+1) \cdot \delta \cdot \mu$. We have

$$H(S_t\xi \vee \dots \vee S_{nt}\xi) \le H(S_t\xi \vee \dots \vee S_{nt}\xi \vee \eta \vee S_\mu\eta \vee \dots \vee S_{k\mu}\eta) = H(\eta \vee S_\mu\eta \vee \dots \vee S_{k\mu}\eta)$$

$$+ H(S_t\xi \vee \dots \vee S_{nt}\xi \mid \eta \vee S_\mu\eta \vee \dots \vee S_{k\mu}\eta) \le H(\eta \vee S_\mu\eta \vee \dots \vee S_{k\mu}\eta)$$

$$+ \sum_{r=1}^{n} H(S_{pt}\xi \mid S_{r(p)\cdot\delta\cdot\mu}\xi)$$

However, $H(S_{pt}\xi \mid S_{r(p)\cdot\delta\cdot\mu}\xi) = H(S_\tau\xi \mid \xi)$, where $\tau = pt - \delta \cdot \mu \cdot r(p) < \delta \cdot \mu$.

We choose an arbitrary $\varepsilon > 0$. Then for sufficiently large m the inequality $H(S_\tau\xi \mid \xi) < \varepsilon$ holds because of the continuity of the flow. Furthermore,

$$H(S_t\xi \vee \dots \vee S_{nt}\xi) \le H(\eta \vee S_\mu\eta \vee \dots \vee S_{k(n)\mu} \cdot \eta) + n\varepsilon .$$

Since $\frac{k(n)}{n} \to \frac{t}{\mu}$ we obtain $h(S_t) \le \frac{t}{\mu} h(S_\mu) + \varepsilon$ in view of the last inequality. But ε is arbitrary, hence $h(S_t) \le \frac{t}{\mu} h(S_n)$.

Now let r be a natural number such that $\frac{t}{r} < \mu$. By Theorem 4 we have $rh(S_{\frac{t}{r}}) = h(S_t)$. By what has been proved, $h(S_\mu) \le \frac{r\cdot\mu}{t} h(S_{\frac{t}{r}})$. Hence $h(S_t) = \frac{t}{\mu} h(S_\mu)$. This proves the theorem.

Lecture 15

THE ENTROPY OF A DYNAMICAL SYSTEM

(Continuation)

The continuity of conditional entropy. Breiman's theorem.

In this lecture, ξ and η will denote measurable partitions of the measurable space $(\mathfrak{M}, \gamma, \mu)$.

By $H(\xi \mid \eta(x))$ we denote the entropy of the partition $\xi \vee \eta$ of the space $(\eta(x), \gamma|_{\eta(x)}, \mu|_{\eta(x)})$, where $\eta(x)$ is the element of the partition η containing the point x.

Exercise. Let $\xi \geq \eta$. Then $H(\xi \mid \eta(x))$ is a measurable function of x.

Definition 1. The conditional entropy is

$$H(\xi \mid \eta) = H(\xi \vee \eta \mid \eta) = \int_{\mathfrak{M}} H(\xi \mid \eta(x))\, d\mu(x) .$$

Exercise. Prove that

$$H(\xi \mid \eta) = - \int_{\mathfrak{M}} \log \mu(\xi(x) \mid \eta(x))\, d\mu(x) .$$

Assertion 1. Let $\{\xi_n\}$ be a sequence of measurable partitions of the space $\mathfrak{M}$ and $H(\xi_1) < \infty$. If $\xi_n \uparrow \xi (\xi_n \downarrow \xi)$ then $H(\xi_n) \uparrow H(\xi)$ $(H(\xi_n) \downarrow H(\xi))$.

Proof. By hypothesis, $\mu(\xi_n(x)) \downarrow \mu(\xi(x)) (\mu(\xi_n(x)) \uparrow \mu(\xi(x)))$ for μ-almost every x. Applying the theorem on the integration of a bounded monotone sequence we obtain the required result.

Assertion 2. Let $\{\xi_n\}$ be as above. Then

$$H(\xi_n|\eta)\uparrow H(\xi|\eta)(H(\xi_n|\eta)\downarrow H(\xi|\eta)) .$$

Proof.

$H(\xi_n|\eta) = \int_{\mathcal{M}|\eta} H(\xi_n|\eta(x)\, d\mu(x)$. By assertion 1, $H(\xi_n|\eta(x))\uparrow H(\xi|\eta(x))$ for all x. It remains to apply the theorem on the integration of a monotone sequence again.

Theorem 1. Let $\{\eta_n\}$ be a sequence of measurable partitions of $\mathcal{M}$, $H(\xi|\eta_1) < \infty$. If $\eta_n \uparrow \eta (\eta_n \downarrow \eta)$ then

$$H(\xi|\eta_n)\downarrow H(\xi|\eta) \quad (H(\xi|\eta_n)\uparrow H(\xi|\eta)) .$$

Proof. Let ξ be a finite partition, $\xi = \{\mathcal{A}_1, \ldots, \mathcal{A}_m\}$, $\mu(\mathcal{A}_k) > 0$. By Doob's theorem, if $i = 1, 2, \ldots, m$ then $\lim_{n\to\infty} \mu(\mathcal{A}_i|\eta_n(x)) = \mu(\mathcal{A}_i|\eta(x))$ holds for μ-almost all x. Consequently, $H(\xi|\eta_n(x)) \to H(\xi|\eta(x))$ for μ-almost all x. Since $H(\xi|\eta(x))$ is a bounded function, from Lebesgue's dominated convergence theorem we obtain

$$H(\xi|\eta_n) = \int_{\mathcal{M}} H(\xi|\eta_n(x))\, d\mu \to \int_{\mathcal{M}} H(\xi|\eta(x))\, d\mu(x) = H(\xi|\eta) .$$

The theorem has been proved for finite partitions ξ. Now let ξ be arbitrary. Then there exists a sequence of finite partitions $\{\xi_n\}$ such that $\xi_n \uparrow \xi$. It remains to apply assertion 2. Theorem 1 has been proved in its entirety.

Let $(\mathcal{M}, \gamma, \mu, T)$ be a dynamical system. Then $h(\xi, T) = H(\xi|\xi^-)$, where $\xi^- = \bigvee_{k=1}^{\infty} T^{-k}\xi$.

Exercise. Let L be a dense subset of Z. If $H(\xi|\alpha) = H(\xi|\beta)$ for all $\xi \in L$ then $\alpha = \beta$.

Exercise. If $H(\xi) < \infty$ and $H(\xi|\eta) = H(\xi)$ then ξ and η are independent.

Theorem 2. If $H(\eta|\xi) < \infty$ and $\eta < \xi_T \overset{\text{def}}{=} \bigvee_{-\infty}^{\infty} T^{-k}\xi$, then $h(\eta, T) \leq h(\xi, T)$.

Proof. Since $\eta < \xi_T$, we have $H(\eta|\xi_T) = 0$. In this way, by Theorem 1, for every $\varepsilon > 0$ there exists s_0 such that $H(\eta|\bigvee_{-s}^{s} T^k\xi) < \varepsilon$ for all $s > s_0$.

For all such s and any n we have

$$\frac{1}{n} H(T\eta \vee \ldots \vee T^n\eta) \leq \frac{1}{n} H(T_\eta \vee \ldots \vee T^n\eta \vee T^{-s}\xi \vee \ldots \vee T^{n+s}\xi) =$$

$$= \frac{1}{n} H(T^{-s}\xi \vee \ldots \vee T^{n+s}\xi) + \frac{1}{n} H(T_\eta \vee \ldots \vee T^n\eta | T^{-s}\xi \vee \ldots \vee T^{n+s}\xi) .$$

But $\frac{1}{n} H(T_\eta \vee \ldots \vee T^n\eta | T^{-s}\xi \vee \ldots \vee T^{n+s}\xi) \leq \frac{1}{n} \sum_{k=1}^{n} H(\eta | T^{-s}\xi \vee \ldots \vee T^s\xi) \leq \varepsilon$

In addition, $\lim_{n\to\infty} \frac{1}{n} H(T^{-s}\xi \vee \ldots \vee T^{n+s}\xi) = h(\xi, T)$ for all s. The assertion of the theorem follows from this.

Remark. The Kolmogoroff generating theorem is a simple consequence of Theorem 2: If ξ is generating for T then $\xi_T = \varepsilon$ and hence $\eta \leq \xi_T$ no matter what the partition η is. If, in addition $\eta \in Z$, then $H(\eta) < \infty$ and, consequently, $H(\eta|\xi) \leq H(\eta) < \infty$. But

$$h(T) = \sup_{\eta \in Z} h(\eta, T) .$$

Now our objective consists in proving the important Breiman theorem.

Theorem 3. Let $(\mathcal{M}, \gamma, \mu, T)$ be an ergodic dynamical system and let $\xi = \{\mathcal{A}_1, \mathcal{A}_2, \ldots, \mathcal{A}_m\}$, $\mu(\mathcal{A}_k) > 0$, $k = 1, \ldots, m$ be a finite partition of the space $\mathcal{M}$. Then for μ-almost every $x \in \mathcal{M}$

$$h(\xi, T) = \lim_{n\to\infty} -\frac{1}{n} \log \mu(\xi_n(x)),$$

where $\xi_n = \xi \vee T^{-1}\xi \vee \ldots \vee T^{-n+1}\xi$.

Before proving the theorem, we will make some preliminary remarks. We denote $\xi^+ = \bigvee_{k=1}^{\infty} T^k \xi$,

$$g_k(x) = -\log \mu(\xi(x) \mid (T\xi \vee \ldots \vee T^k\xi)(x)),$$

$$k = 1, 2, \ldots$$

$$g_0(x) = -\log \mu(\xi(x)), \quad g(x) = -\log \mu(\xi(x) \mid \xi^+(x)) .$$

(By Doob's theorem, $\lim_{k\to\infty} g_k(x) = g(x)$ μ- a.e.).

Lemma. $\int_{\mathcal{M}} [\sup g_k(x)]\, d\mu(x) < \infty$.

Proof of Lemma. It suffices to prove that there exists a constant C such that $\mu\{x : \sup_k g_k(x) > \lambda\} \leq Ce^{-\lambda}$ for all λ. Let $f_k^i = -\log \mu(\mathcal{A}_i \mid (T\xi \vee \ldots \vee T^k\xi)(x))$ $E_1 = \{x : g_i(x) > \lambda\}$, $F_1^i = \{x : f_1^i(x) > \lambda\}$. For $k \geq 2$ we set

$$E_k = \{x : \max_{j<k} g_j(x) < \lambda,\ g_k(x) > \lambda\},$$

$$F_k^i = \{x : \max_{j<k} f_j^i(x) \leq \lambda,\ f_k^i(x) > \lambda\} .$$

We have $\mu(E_k) = \sum_{i=1}^{m} \mu(E_k \cap \mathcal{A}_i) = \sum_{i=1}^{m} \mu(F_k^i \cap \mathcal{A}_i)$. Since $F_k^i \in \gamma(\xi \vee T\xi \vee \ldots \vee T^k\xi)$, the theorem on conditional probability yields

$$\mu(\mathcal{A}_i \cap F_k^i) = \int_{F_k^i} \mu(\mathcal{A}_i | (T\xi \vee \ldots \vee T^k\xi)(x))\, d\mu(x) = \int_{F_k^i} e^{-f_k^i(x)}\, d\mu(x) \leq e^{-\lambda}\mu(F_k^i) .$$

If $k \neq j$, then $F_k^k \cap F_j^i = \phi$, hence

$$\mu\{x : \sup_k g_k(x) > \lambda\} = \sum_k \mu(E_k) \leq \sum_{i=1}^{m} e^{-\lambda} \sum_k \mu(F_k^i) \leq me^{-\lambda} .$$

The lemma has been proved.

Proof of Breiman's theorem. It is easy to verify that

$$-\frac{1}{n}\log \mu(\xi_n(x)) = \frac{1}{n}\sum_{k=0}^{n-1} g_k(T^k(x)) = \frac{1}{n}\sum_{k=0}^{n-1} g(T^k x)$$

$$+ \frac{1}{n}\sum_{k=0}^{n-1} [g_k(T^k x) - g(T^k x)] .$$

By the ergodic theorem of Birkhoff-Khinchin, the first sum converges to $\int g(x)\, d\mu(x) = h(\xi, T)$. In this way, in order to prove the theorem, it is necessary to prove

$$\lim \frac{1}{n}\sum_{k=0}^{n-1} (g_k(T^k x) - g(T^k x)) = 0 \qquad (*)$$

We set $G_N(x) = \sup_{k \geq N} |g_k(x) - g(x)|$. We have

$$\left|\frac{1}{n}\sum_{k=0}^{n-1} (g_k(T^k x) - g(T^k x))\right| \leq \frac{1}{n}\sum_{k=0}^{n-1} |g_k(T^k x) - g(T^k x)|$$

$$= \frac{1}{n}\sum_{k=0}^{N-1} |g_k(T^k x) - g(T^k x)| + \frac{1}{n}\sum_{k=N}^{n-1} |g_k(T^k x) - g(T^k x)| \leq \frac{1}{n}\sum_{k=0}^{N-1} |g_k(T^k x) -$$

$$- g(T^k x)| + \frac{1}{n}\sum_{k=N}^{n-1} G_N(T^k x) .$$

If $n \to \infty$ in the previous inequality and N is fixed, then

$$\lim_{n\to\infty} \left|\frac{1}{n} \sum_{k=0}^{n-1} (g_k(T^k x) - g(T^k x))\right| \leq \lim_{n\to\infty} \frac{1}{n} \sum_{k=N}^{n-1} G_N(T^k x) = \int_{\mathcal{M}} G_N(x)\, d\mu(x)$$

for μ-almost all x (by Birkhoff-Khinchin theorem). Furthermore, $0 \leq G_N(x) \leq$ $\leq g(x) + \sup_k g_k(x)$. By the lemma, $g(x) + \sup_k g_k(x)$ is an integrable function. In this way,

$$\lim_{N\to\infty} \int_{\mathcal{M}} G_N(x)\, d\mu(x) = \int_{\mathcal{M}} \lim_{N\to\infty} G_N(x)\, d\mu = 0$$

Equality (*) and the Breiman theorem are proved with this.

Examples of computation of entropy.

The entropy of an automorphism of the torus.

For the computation of the entropy of a group automorphism of the two-dimensional torus we will make use of the Markov partition that we constructed in Lecture No. 12.

Recall that, for the ergodic automorphism T of the two-dimensional torus, given by the integral matrix $\begin{pmatrix} a & b \\ c & d \end{pmatrix}$ with determinant equal to one, a Markov partition into parallelograms $\xi = (C_1, C_2, \ldots, C_r)$ was constructed. As was $h(T) = h(T, \xi)$ by the theorem of Kolmogorov. We will occupy ourselves with the computation of $h(T, \xi)$. We will show that, for every set $C_{i_0} \cap T^{-1} C_{i_1} \cap \ldots \cap T^{-n} C_{i_{-n}}$, the following relation takes place:

$$a_2 \leq \frac{\mu(C_{i_0} \cap T^{-1} C_{i_{-1}} \cap \ldots \cap T^{-n} C_{i_{-n}})}{\lambda^n} \leq a_1, \tag{1}$$

if this intersection has positive measure, a_1, a_2 are constants independent of n, and λ is an eigenvalue of the matrix $\begin{pmatrix} a & b \\ c & d \end{pmatrix}$ with modulus less than one.

It is clear from this that the number of elements of the partition $\xi \vee T\xi \vee \ldots \vee T^n\xi$ does not exceed $a_i \lambda^{-n}$ and $H(\xi \vee T^{-1}\xi \vee \ldots \vee T^{-n}\xi) =$
$= - \Sigma\mu(C_{i_0} \cap T^{-1}C_{i_{-1}} \cap \ldots \cap T^{-n}C_{i_{-n}}) \ln \mu(C_{i_0} \cap \ldots \cap T^{-n}C_{i_{-n}}) \leq \ln a_1 - n \ln \lambda.$
We obtain the estimate from below in an analogous way. It follows easily from these estimates that $h(T, \xi) = - \ln \lambda$. For the proof of inequalities (t) we remark that, by virtue of the properties of a Markov partition, the intersection

$$C_{i_0} \cap T^{-1}C_{i_{-1}} \cap \ldots \cap T^{-n}C_{i_{-n}}$$

is a parallelogram whose sides are parallel to the sides of the parallelogram C_{i_0}. Wence the length of the contracting side of this parallelogram is equal to the length of the contracting side of the parallelogram C_{i_0}, and the length of the expanding side is equal to the product of λ^n by the length of the expanding side of the parallelogram $C_{i_{-n}}$. The required inequality follows from this.

<u>The entropy of an ergodic shift on a compact commutative group.</u>

We begin with the following observation: If for some partition ξ and some $a > 0$ the number of distinct elements of nonzero measure of the partition

$$\xi \vee T^{-1}\xi \vee \ldots \vee T^{-n}\xi$$

does not exceed n^a for sufficiently large n then

$$h(T, \xi) = 0 .$$

For the proof we arrange the measures of the elements of the partition in decreasing order. We obtain a sequence of positive numbers $\pi_1 \geq \pi_2 \geq \geq \dots \geq \pi_k$, where $k \leq n^a$, $\Sigma \pi_k = 1$. We find a number i such that $\pi_i \geq \frac{1}{n^{4a}} > \pi_{i+1}$. If $\pi_k > \frac{1}{n^{4a}}$ then we set $i = k$. Then

$$- \sum_{s=1}^{k} \pi_s \ln \pi_s = - \sum_{s=1}^{i} \pi_s \ln \pi_s - \sum_{s=i+1}^{k} \pi_s \ln \pi_s = I_1 + I_2 .$$

For I_1 we have the following estimate:

$I_1 \leq 4a \ln n$ and hence $\frac{I_1}{n} \to 0$, $n \to \infty$.

As far as I_2 is concerned, we observe that $-\mu \ln \mu \leq \mu^{\frac{1}{2}}$ for small μ. Hence

$$I_2 = - \sum_{i+1}^{k} \pi_s \ln \pi_s \leq \sum_{i+1}^{k} \pi_s^{\frac{1}{2}} \leq n^a \cdot n^{a - \frac{4}{2}} \leq 1.$$

Consequently, $\frac{I_2}{n} \to 0$, $n \to \infty$, and our assertion is proved.

We will examine the rotation of the circle through an irrational angle a. We will prove that its entropy is equal to zero. Let the partition ξ of our circle consist of two semicircles. We will prove that this partition is generating. Indeed, let x and y be distinct points on the circle. Then the point z which is either of the two points determining the partition ξ, will appear between x and y after a sufficiently high number of shifts (by virtue of the density of its trajectory). Hence at some step they will be in different elements of the partition $\xi \vee T^{-1}\xi \vee \dots \vee T^{-n}\xi$. This means that ξ is a generating partition.

Now we observe that the number of elements of the partition $\xi \vee \dots \vee T^{-n}\xi$ with nonzero measure grows linearly with n. Indeed, after each step it increases by precisely two elements. Hence the entropy of the shift is equal to zero.

The example of shifts of the two-dimensional torus is examined in exactly the same way. Using some more deep arguments one can show that the entropy of any shift on any compact abelian group is equal to zero.

The entropy of a flow over a two-dimensional torus.

Let $\mathcal{M}$ be a two-dimensional torus and let S_t be a flow on $\mathcal{M}$ generated by the system of differential equations

$$\begin{aligned} \frac{dx}{dt} &= F_1(x, y) \\ \frac{dy}{dt} &= F_2(x, y) \end{aligned} \tag{1}$$

preserving the invariant measure $d\mu = \rho(x, y)\, dx\, dy$.

We will assume that F_1, F_2 and ρ are functions of class C^∞ in order to avoid small complications connected with finite differentiability. Furthermore, we will assume that $F_1 > 0$. As was shown in Lecture 11, it is always possible to introduce smooth coordinates on the torus such that the trajectories of our system are transformed into straight lines $y = \alpha x + c$. This is equivalent to the assertion that, in an appropriate coordinate system, the system of differential equations (1) becomes the system of differential equations (2):

$$\begin{aligned} \frac{dx}{dt} &= f(x, y) \\ \frac{dy}{dt} &= \alpha f(x, y) \end{aligned} \tag{2}$$

where $f(x, y)$ is a function of class C^∞ that does not vanish.

In this example it is convenient to become acquainted with a construction that is often used in ergodic theory.

Let Q be a space with a normalized measure σ, and let T be a measurable, measure preserving transformation of Q. Let F be a measurable, positive almost everywhere, and integrable with respect to the measure σ, function defined on Q.

We will examine the subset $\mathcal{M}$ of the product $R^1 \times Q$:

$$\mathcal{M} = \{(q,\mu) : q \in Q,\ 0 \le \mu \le F(q)\}$$

The structure of the product $R^1 \times Q$ induces a measure ν on the space $\mathcal{M}$ which is not normalized in general: $\nu(\mathcal{M}) = \int_Q F(q)d\sigma(q)$. We can normalize this measure. The measure obtained in this way will be denoted μ.

We will examine the one parameter group of transformations of the space $\mathcal{M}$ which acts in the following way: A point q_0 moves from the lower level with unit speed vertically upwards in time $F(q_0)$, and at the moment $F(q_0)$ it reaches the point Tq_0, and so forth. In this way we obtain a one parameter group of transformations of the space $\mathcal{M}$. It is easy to prove that it preserves the measure μ (for piece-wise constant functions F this is obvious, and the standard approximation is performed for arbitrary F). A dynamical system is produced in this way. It is called the special flow constructed with the automorphism T of the space Q and the function F.

An important theorem of ergodic theory (Ambrose-Kakutani theorem) asserts that under very general conditions every dynamical system is isomorphic in the sense of ergodic theory to some special flow.

Remark 1. The idea of reducing the study of systems of differential equations with continuous time to the study of iterations of individual transformations goes back to Poincaré (the cross-section method).

Remark 2. The idea of the special flow is also interesting from the following point of view. The space Q, contained in $\mathfrak{M}$, is seen to have measure zero. But every trajectory of our flow passes through this set. In this way, the possibility appears of distinguishing in a phase space essential sets of measure zero as sets through which a set of trajectories of positive measure passes. This concept is useful in the investigation of certain infinite dimensional dynamical systems.

Actually, we have already met with special flows several times in the representations of certain dynamical systems. For example, we will examine billiards in a closed region Q with boundary Γ. The phase space consists of the linear elements whose carriers lie in the region Q. We take, in the capacity of Q, the linear elements whose carriers lie in Γ and which are directed into Q. Here the transformation T arises naturally as the result of the first reflection of the trajectories from the boundary Γ.

Another example relates to the dynamical system on the two-dimensional torus. Let Γ be the circle $\Gamma = \{x = 0\}$. The trajectory starting from the point $(0, y)$ will return after some time to some point of the same circle, which we denote $(0, Ty)$. In this way we obtain a transformation T of the circle Γ induced by our dynamical system. We denote by $\tau(y)$ the time of motion of the point y to the first time it hits the circle Γ. The system (2) is none other than the special flow on the rotation of the circle through an angle α with the function F equal to τ.

Now we will formulate (without proof) the important theorem of L. M. Abromov:

Let Q be a Lebesgue space with a normalized measure μ. Let T be an automorphism of Q and let F be a measurable function on Q; also $F \geq \tau > 0$ and let $\{S_t\}$ be the special flow constructed with the automorphism T and the function F.

Then $h(S_t) = |t| \dfrac{h(T)}{\int_Q F(q)\,d\mu(q)}$.

Applying this theorem to the dynamical system on the two-dimensional torus we obtain $h(S_t) = 0$ because $h(T) = 0$ in view of the fact that T is a rotation of the circle.

THE ENTROPY OF BILLIARDS INSIDE A POLYGON

Theorem. The entropy of billiards inside a polygon is equal to zero.

Proof: We begin with a simple remark. Let α be a finite partition with elements $\mathcal{A}_1, \ldots, \mathcal{A}_N$ of a measure space M and T be a measure preserving transformation of this space. Assume that it is possible to find for any $\varepsilon > 0$ a number $n_0(\varepsilon)$ and a set $\mathcal{M}_1 \subseteq \mathcal{M}$ such that $\mu(\mathcal{M}_1) < \varepsilon$ and for any $n \geq n_0(\varepsilon)$ from $\mathcal{A}_{i_0} \cap T\mathcal{A}_{i_1} \cap \ldots \cap T^n \mathcal{A}_{i_n} \cap (\mathcal{M} - \mathcal{M}_1) \neq \phi$ it follows that

$$\mu(\mathcal{A}_{i_0} \cap T\mathcal{A}_{i_1} \cap \ldots \cap T^n \mathcal{A}_{i_n}) \geq c(\varepsilon) n^{-\gamma} \qquad (*)$$

where $\gamma > 0$ is some absolute constant and $c(\varepsilon)$ depends only on ε. In this case $h(T, \alpha) = 0$.

Indeed, one can write

$$H(\alpha \vee T\alpha \vee \ldots \vee T^n \alpha) \leq H(\alpha \vee T\alpha \vee \ldots \vee T^n \alpha \vee \beta) = H(\beta) + H(\alpha \vee T\alpha \vee \ldots \vee T^n \alpha \mid \beta)$$

where β is the partition of $\mathcal{M}$ into two sets $\mathcal{M}_1$ and $\mathcal{M} - \mathcal{M}_1$. From the definition of the conditional entropy

$$H(\alpha \vee T\alpha \vee \ldots \vee T^n \alpha \mid \beta) =$$

$$-\mu(\mathcal{M}_1) \Sigma \mu(\mathcal{A}_{i_0} \cap \ldots \cap T^n \mathcal{A}_{i_n} \mid \mathcal{M}_1) \cdot \ln \mu(\mathcal{A}_{i_0} \cap \ldots \cap T^n \mathcal{A}_{i_n} \mid \mathcal{M}_1)$$

$$-\mu(\mathcal{M} - \mathcal{M}_1) \Sigma \mu(\mathcal{A}_{i_0} \cap \ldots \cap T^n \mathcal{A}_{i_n} \mid \mathcal{M} - \mathcal{M}_1) \ln \mu(\mathcal{A}_{i_0} \cap \ldots \cap T^n \mathcal{A}_{i_n} \mid \mathcal{M} - \mathcal{M}_1)$$

The first term is no more than $\varepsilon \ln N \cdot n$. From (*) it follows that for any nonzero summand in the second term $-\ln \mu(\mathcal{A}_{i_0} \cap \ldots \cap T^n \mathcal{A}_{i_n} | \mathcal{M} - \mathcal{M}_1) \leq \gamma \ln n - \ln C(\varepsilon)$ and therefore

$$-\mu(\mathcal{M}-\mathcal{M}_1) \Sigma \mu(\mathcal{A}_{i_0} \cap \ldots \cap T^n \mathcal{A}_{i_n} | \mathcal{M}-\mathcal{M}_1) \ln \mu(\mathcal{A}_{i_0} \cap \ldots \cap T^n \mathcal{A}_{i_n} | \mathcal{M}-\mathcal{M}_1) \leq \gamma \ln n - \ln C(\varepsilon)$$

$$\frac{1}{n} H(\alpha \vee \ldots \vee T^n \alpha) \leq \varepsilon \ln N + \frac{\gamma \ln n}{n} - \frac{\ln C(\varepsilon)}{n} + \frac{1}{n} H(\beta) .$$

Thus, $h(T,\alpha) \leq \varepsilon \ln N$ and $h(T,\alpha) = 0$ because ε is arbitrary.

Now let us return to our billiards system. We shall deal with the transformation T which appears in the natural special representation of our billiards (see the precedent page). Let Q be a domain on the plane the boundary of which is a polygon Γ. We shall take as $\mathcal{M}$ the space of line elements $x = (q, v)$ whose carries lie in Γ and which are directed into Q. It is easy to see that the invariant measure for T takes the form

$$d\mu = \cos \varphi \, dq \, d\varphi,$$

where φ is the angle between v and the interior normal vector to Γ in q, $-\frac{\pi}{2} < \varphi < \frac{\pi}{2}$.

We shall consider the following partitions α of $\mathcal{M}$. Let Γ be partitioned into k intervals Δ_i, $i=1,\ldots,k$ and the interval $-\frac{\pi}{2} < \varphi < \frac{\pi}{2}$ be partitioned into ℓ intervals Δ'_j, $j=1,\ldots,\ell$. We define α as a partition the elements $\mathcal{A}_{ij}$ of which are of the form $\mathcal{A}_{ij} = \{x=(q,v): q \in \Delta_i, v \in \Delta'_j\}$. It is sufficient to prove that $h(T,\alpha) = 0$ for any such α. We shall denote by $\partial \mathcal{A}_{ij}$ the boundary of $\mathcal{A}_{ij}$ and by dist a natural metric in $\mathcal{M}$.

Lemma 1. Let $B_n = \{x \in \mathcal{M}: \text{dist}(x, \partial \mathcal{A}_{ij}) \leq \frac{1}{n^2}, x \in \mathcal{A}_{ij}\}$. Then $\mu(B_n) \leq \frac{\text{const}}{n^2}$

It is obvious.

Lemma 2. Let $C_n = \{x \in \mathcal{M} : T^n x \in B_n\}$. Then $\mu(C_n) \leq \frac{\text{const}}{n^2}$.

Obvious because $\mu(C_n) = \mu(B_n)$. Thus we have $\Sigma \mu(C_n) < \infty$. It follows from Borel-Cantelli lemma that for any $\varepsilon > 0$ one can find a number $K = K(\varepsilon)$ and a set $\mathcal{M}_1$ such that $\mu(\mathcal{M}_1) < \varepsilon$ and for any $x \in \mathcal{M} - \mathcal{M}_1$ $\text{dist}(T^n x, \mathcal{A}_{ij}) \geq \frac{K(\varepsilon)}{n^2}$, $T^n x \in \mathcal{A}_{ij}$ for all $n \geq 0$. Let $x \in \mathcal{M} - \mathcal{M}_1$ and $\mathcal{U}_n(x)$ be the ball of radius $L \cdot n^{-6}$ with centre in x, where L is a constant which will be defined below. We shall show that $\mathcal{U}_n(x)$ lies in a single element of the partition $\alpha \vee T\alpha \vee \ldots \vee T^n \alpha$. If this is shown then $\mu(D_n) \geq \mu(\mathcal{U}_n(x)) \geq \frac{\text{const } L^2}{n^{12}}$ where $D_n \geq \mathcal{U}_n(x)$ is the element of $\alpha \vee \ldots \vee T^n \alpha$ containing x. In other words, $-\ln \mu(D_n) \leq 12 \ln n - 2 \ln L - \ln(\text{const})$ and for sufficiently large n the inequality (*) is true. Therefore $h(T, \alpha) = 0$. From this it follows easily that $h(T) = 0$.

Now we are going to prove that $\mathcal{U}_n(x) \subset D_n$ where D_n is the element of $\alpha \vee T\alpha \vee \ldots \vee T^n \alpha$ containing x, $x \in \mathcal{M} - \mathcal{M}_1$. Having the point $x = (q, v)$, let us construct a ray on the plane taking x as the beginning. We shall use the method of

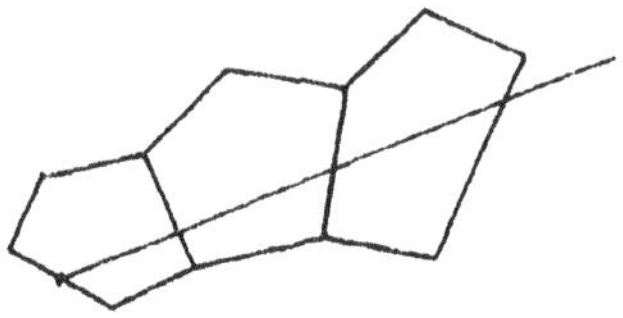

reflection. It is necessary to stress that this method can be applied only for reflections from polygons. We shall reflect our polygons along the ray in such a way that a natural mapping of these polygons onto the initial one gives a trajectory of the point x inside Q.

This mapping is a isometry. It follows from this fact that if we consider for some t and $\delta > 0$ a ball $W_\delta(x)$ of radius δ with x as the centre and construct the ray of the length t beginning at $x' \in W_\delta(x)$ and if each and any such ray intersects the same sides of polygon as the ray going out of x, then under our isometry it transforms into the interval of the billiard trajectory of the length t, beginning at x'.

Now assume that $x \in \mathfrak{M} - \mathfrak{M}'$ and for any n let us take t such that $F(x) + F(Tx) + \ldots + F(T^n x) \geq t$. Here F is the function entering in the special representation of our billiard system. It is easy to see that $t \leq \text{const}\, n$. Putting $\delta = \frac{K(\varepsilon)}{\text{const}\, n^3}$ we shall have for each $x' \in W_\delta(x)$ that the points of intersection of the interval of the length t for the trajectory beginning at x' lie in the same element of the partition α as the corresponding point of the intersection for the trajectory beginning at x. Thus, our assertion is proved.

References

1. P. Billingsley, Ergodic Theory and Information, Wiley, New York, 1965.

2. V. I. Arnold, A. Avez, Ergodic Problems of Classical Mechanics, Benjamin, New York, 1968.

3. K. Jacobs, Lecture Notes on Ergodic Theory, vol. 1, 2, Univ. of Aarkus, Denmark, 1962-1963.

4. M. Smorodinsky, Ergodic Theory. Entropy

5. Ja. G. Sinai, Dynamical Systems, Part I, Aarhus University, Denmark.

6. V. A. Rohlin, Lectures on entropy theory of measure-preserving transformations, Uspehi Math. Nauk, 1967, n. 22, n5, 3-56.

7. D. Ornstein, Ergodic Theory, Randomness and Dynamical Systems, Yale Math. Monographs, n5.

Milton Keynes UK
Ingram Content Group UK Ltd.
UKHW012232081223
434043UK00001B/78

9 780691 081823